Self-Compacting Concrete

Self-Compacting Concrete

Edited by
Ahmed Loukili

First published 2011 in Great Britain and the United States by ISTE Ltd and John Wiley & Sons, Inc.

ISTE Ltd
27-37 St George's Road
London SW19 4EU
UK

www.iste.co.uk

John Wiley & Sons, Inc.
111 River Street
Hoboken, NJ 07030
USA

www.wiley.com

© ISTE Ltd 2011

Library of Congress Cataloging-in-Publication Data

Self-compacting concrete / edited by Ahmed Loukili.
    p. cm.
  Includes bibliographical references and index.
  ISBN 978-1-84821-290-9
  1. Self-consolidating concrete. I. Loukili, Ahmed.
  TA442.5.S45 2011
  620.1'36--dc23
                                                                    2011020213

British Library Cataloguing-in-Publication Data
A CIP record for this book is available from the British Library
ISBN 978-1-84821-290-9

Printed and bound in Great Britain by CPI Antony Rowe, Chippenham and Eastbourne.

Table of Contents

Introduction

Self-compacting concretes (SCCs), highly fluid concretes placed without vibration, were introduced into French construction works towards the end of the 1990s. The concept came into being a decade earlier in Prof. Okumara's laboratory [OKA 00] in Japan. The high seismicity of this geographical region requires the use of high levels of steel reinforcement in construction. The use of "self-compacting" concretes appeared as a solution to improve the filling up of zones which are not very accessible to conventional methods of concrete compaction. This solution also has the advantage of overcoming the gradual decline in the number of workers qualified to handle and place concrete.

In France, SCC was initially of interest to the precast concrete and ready mix concrete industries, and in the construction industry, well before project managers and contracting authorities became interested in it [CIM 03]. The use of SCC enables improvements in productivity through reductions in manpower and placing delays. It also improves quality through a better filling of the formwork, better coating of the steel reinforcement, even a better facing. Finally, and undeniably their best asset, SCCs reduce the difficulty of the work. By preventing vibration, the health effects of concrete construction disappear (white hand syndrome, hearing loss, noise disturbances for the neighbors). Little by little, SCCs have also won over architects by offering them the possibility of playing with complex volumes.

Even though SCCs have established their position in the prefabrication industry (around half of the volume produced), SCCs used *in situ* are struggling to make an impact on construction sites, in France as well as in other countries [SHA 07]. Despite their numerous advantages, SCCs represent less than 3% of ready mix concrete produced in France [BTP 07]. Several factors lend themselves to explaining this slow expansion of SCCs [CUS 07]. Firstly, making SCCs is somewhat difficult, since the components must be of a good quality and have little variation in their properties. While the properties of fresh vibrated concretes are affected relatively little by normal variations in the components (size distribution, water content, etc.), SCCs, on the other hand, are much more sensitive. Secondly, the production tool is not always precise enough for making concretes which are strongly affected by errors in the mixture proportions. Thirdly, the formworks must be well prepared, properly waterproofed and must, above all, be able to withstand pressures that are *a priori* higher than those involved in handling vibrated concretes.

However, SCCs have the potential of continuing to expand. To begin with, the standardizing framework, which had previously been vague in Europe, was enforced in June 2010 with the release of the EN 206-9 standard which brought in rules for production, handling, and specific controls for SCC, complementing EN 206-1. SCCs are becoming widespread elsewhere by strengthening the dialog – which is truly indispensable – between construction agents, owners, project managers, architects, businesses and suppliers, and also research laboratories. SCCs, complex and innovating materials, have been the object of a real infatuation by researchers the world over. As a witness to this success, international conferences have been dedicated to SCCs since 1999 [SCC 99]. Today the extent of the research allows us to have a better understanding of the behavior of these concretes.

The objective of this book is therefore to disseminate knowledge acquired by recent research in order to enable the student, the technician, or the engineer who reads it, to develop an understanding of the formulation of these materials. The composition of SCCs must satisfy several criteria. In addition, different authors have endeavored to reply to each of the questions posed in the following chapters, without losing sight of the global objective of techno-economical optimization.

Chapter 1 is dedicated to rheology and concrete casting. Theoretical concepts are presented and useful experimental tools for characterizing the behavior of these complex mixtures are described. Experimental data also shows the range of variability and the influence of the principal formulation parameters.

Chapter 2 enables the reader to understand the specifics of the behavior of SCCs at early ages. This behavior, which is strongly influenced by the particular formulations of SCCs, is characterized by vulnerability to desiccation and the resultant strains.

Chapter 3 focuses on the mechanical and delayed behaviors of SCCs in comparison with ordinary derivative concretes. This aspect is crucial for designing self-compacting concrete pieces.

In Chapter 4, the question of durability is examined. Degradation phenomena linked to environmental events are described, and experimental data on SCC and vibrated concretes are brought together to show which parameters are influential from the point of view of potential durability.

Finally Chapter 5 is dedicated to the thermal stability and fire resistance of self-compacting concretes.

Bibliography

[CIM 03] "CimBéton, Monographie d'ouvrages en BAP", *Collection Technique Cimbéton*, B 52, 2003.

[CUS 07] CUSSIGH F., "SCC in practice: opportunities and bottleneck", *Proceedings of the Fifth RILEM Symposium on Self-Compacting Concrete*, Ghent, Belgium, 2007.

[FRA 07] France BTP.com, "Le BAP: où en est-on en 2007?", *BTP Matériaux*, December 2007.

[OKA 00] OKAMURA H., OZAWA K., OUCHI M., "Self-Compacting Concrete", *Structural Concrete*, vol. 1, no. 3, 2000.

[SCC 99] *Proceedings of the First International RILEM Symposium on Self-Compacting Concrete*, Stockholm, Sweden, 1999.

[SHA 07] SHAH S.P., FERRON R.P., FERRARA L., TREGGER N., KWON S.H., "Research on SCC: some emerging themes", *Proceedings of the Fifth RILEM Symposium on Self-Compacting Concrete*, Ghent, Belgium, 2007.

Chapter 1

Design, Rheology and Casting of Self-Compacting Concretes

1.1. Towards a fluid concrete

Recent decades have witnessed a remarkable evolution in concrete performance, as much in the field of their rheological behavior in the fresh state as in their mechanical behavior in their hardened state. These technical advances are the results of coupling between the formulation principles, coming from a long period of learning by experience and the mastering of physico-chemical principles which govern the behavior of cement-based materials.

Initially made using a simple recipe of water, cement and aggregates (sand and gravel), concrete has since seen its formulation enriched by the inclusion of high quality components such as mineral additives (limestone fillers,

Chapter written by Sofiane AMZIANE, Christophe LANOS and Michel MOURET.

silica fumes, etc.), chemical additives such as super-plasticizers and reinforcing materials such as fibers.

The complex formulations thus obtained must satisfy the production specifications in which the obligations often go beyond the conventional output requirements in terms of fluidity during casting, and strength of the hardened concrete. Concrete design can thus be adjusted to suit the working conditions (pumping, vibration, transport time and casting time), hardening (time at which the concrete is removed from the mold, required short-term strength) and service (developing strength, durability, etc.).

Self-compacting concrete (SCC) is an illustration of research in mastering such complex mixtures. The origin of SCCs is associated with the development, at the beginning of the 1980s in Japan, of a design method for fluid concretes. The high seismicity of this geographical region necessitates that structures are highly reinforced with steel. In these much more difficult pouring conditions, compacting the resultant concrete using conventional methods (internal or external vibration) is at risk of being insufficient, thereby compromising the buildings' quality assurance.

The SCC concept was therefore born from the desire to make the concrete compacting completely independent of the production context, whether in the technical plan or in manpower, knowing that the number of qualified workers is noticeably declining in Japan, which is also the case in numerous other countries.

SCC is therefore a mixture which is both fluid and homogenous, which fills formworks perfectly by flowing under the effect of gravity alone, and which completely wraps around all the reinforcing bars without causing blockages or grain separation in their vicinity (Figures 1.1 and 1.2). These properties appeal to the structure designer for the work, who can envision more complicated shapes using these materials; and to the production team who is interested in simplifying the work involved in creating the structures and reducing construction delays. This is why the use of SCCs tends to increase on a large scale.

Figure 1.1. *Filling capability of SCC (top) and conventional concrete (bottom) [CAS 05]. SCC, moving under its own weight, flows around obstacles and ensures good coverage. Without an external energy supply, such as vibration, conventional concrete cannot achieve the same result*

On the construction site, the use of SCC results in the interruption of vibration techniques leading to significant reductions in noise and occupational illnesses such as white hand syndrome – Raynaud's syndrome – improvements in safety during casting and reduction in the workforce required for casting. It also improves the conditions for filling formworks by increasing the concreting speed, allowing concreting in highly reinforced zones, faster formwork rotation. Finally, the use of SCCs results in a better facing quality (Figure 1.2).

Figure 1.2. *Improvements in facing quality with SCC [CAS 05]. Casting is made easier with SCC which reduces the formation of air bubbles and pebble clusters at the faces of the formwork, including frames with complex geometrical shapes. In the case of conventional concrete, such defects can be eliminated using vibration which is always difficult to implement*

This collection of properties affects construction as much as project managers do, since they are also involved in improving productivity during the construction phase, in the quality of constructions and in their durability.

The transition to SCCs cannot be envisaged without reconsidering and adapting the fabrication and casting methods. Numerous works have therefore been committed to the problems of mixing, transport, pumpability, formworks (waterproofing and controlling pressure on the inner surfaces), facing quality, etc., according to the workable nature of SCCs. Controlling the production of SCCs, relying on simple and rudimentary tests in construction sites also constitutes a theme for collective consideration which has led to the publication of a reference document in France.

Recommandations pour l'emploi des bétons auto-plaçants [AFG 08], classifies SCCs, specifies the required properties as well as their corresponding tests, and describes the conditions of use and the characteristics of these concretes. However, operators confronted with formulating, producing or casting SCC have systematically noticed that this type of mixture is characterized by a fluidity which is strongly affected by all deviations in composition. To understand the reasons behind this, the design basics are initially discussed.

The various arguments dedicated to developing an understanding of a mixture's flow properties with respect to quantifiable and representative rheological parameters are then introduced. Next, the most suitable test methods for estimating these rheological parameters are presented. The influence of design parameters on the values of the rheological parameters is also studied. The different arguments make it possible to make judgments of a formulation's robustness in light of measurement differences which are often the origin of a lack of reproducability. Armed with this knowledge of design principles and rheological parameters to consider, industrial practices regarding mixing, transport and pumping are developed. Finally the pressure exerted by SCCs, once in place, on formwork is described in the results from recent works.

1.1.1. *Area of application*

At the time of writing, SCCs represent 2-3% of the volume of ready mix concrete produced in France, with a tendency towards horizontal applications (80%) by comparison with vertical applications (20%). A 40-50% proportion of SCC is intended to precast products with delayed demolding [MAG 07].

Whether we are concerned with horizontal structures (floors, raft foundations, paving, screed, etc.) or with vertical structures (columns, walls, bridge piers, etc.), the function of SCCs in their fresh and hardened states are singularly different.

For horizontal structures, the term self-leveling concrete (SLC) is sometimes used. In their fresh state, finishing quality and leveling is crucial. This is obtained by putting concrete in place without recourse to mechanical work, especially for the surface. Neither segregation nor excessive bleeding is tolerated. For this type of application, the current method combines thickening agents (starches, *Welan gum,* etc.) with super-plasticizers to achieve the ideal fluidity/stability coupling. In the hardened state, the strength is of the order of 40 MPa.

For vertical structures, the generic term self-compacting concrete (SCC) is used. Vertical elements are often load-bearing, which requires a high level of steel-reinforcement and a high concrete strength to reduce the section of pieces installed. Targeted strengths can reach the limits of Eurocode 2, i.e. 90 MPa. The difficulty in formulating an SCC with high mechanical performance stems from the conflict between the mechanical performance objective and that of fluidity and stability. The first objective is linked to optimizing the cement hydration reaction for a water/cement mass ratio of 0.23 (stochiometric optimum). The second objective leads straightforwardly to a water/cement mass ratio of 0.6. In order to guarantee high strength, reactive mineral additives and super-plasticizers which effectively reduce the water requirements must be used.

Whatever the intended application may be, the use of specific chemical additives and large proportions of mineral admixtures must be understood, in a different way to conventional approaches to concrete composition [BAR 97], formulating cement + mineral additive pairs and determining the amount of aggregate in the concrete.

1.2. SCC formulation basics

1.2.1. *Overview*

Formulating SCCs is a compromise between sufficiently high fluidity to ensure good casting and an adequate consistency to avoid phase separation problems (segregation or bleeding).

The principal idea in designing a SCC involves considering the mixture as a concentrated suspension in which the suspending phase is viscous and dense enough and in which the dispersed phase is at a low enough concentration to prevent too much interaction. The suspending phase is often called "the paste". It comprises finely-sized components, chemical additives and water. The dispersed phase includes larger-sized components (aggregates). According to this simplistic explanation, the separation risk is therefore directly linked to the difference in densities between the aggregates and the paste. The volumetric proportion of the paste in the mixture must be high enough to limit interactions and thereby reduce the risk of blockages during pouring. Consequently, the overall fluidity of the mixture turns out to be strongly linked to the quantity and fluidity of the paste.

The definition of a clear boundary between the paste and occlusions in terms of particle size is, however, difficult and arbitrary. Conventional methods of formulating vibrated concretes generally consider that fine particles are particles with a diameter of less than 63 µm, or 80 µm according to the authors. This boundary reflects a change in the type of interaction between the particles. The largest particles interact mainly by contact and friction between the grains. The interactions of the finest particles are modified by colloidal effects, for example, these effects are particularly affected by introducing additives. In the case of SCCs, it is important to consider that this boundary can be modified as a function of the flow conditions. In static phases, the distinction between the paste and aggregate goes back to a small particle size whereas during a significant shearing of the concrete, the boundary is moved back towards the largest size of aggregates.

Moving these basic SCC formulations back in the form of ratios between the components therefore turns out to be complicated. It is possible, however, to resolve this problem by referring to the basic formulations of conventional vibrated concretes. Figure 1.3 shows schematically how the formulation of an SCC differs from that of a conventional vibrated concrete [TUR 04]. At similar levels of cement, sand and water, SCC yields a greater quantity of paste through the addition of mineral admixtures and a lower quantity of gravel. The volumetric increase in the paste, associated with a reduction in the quantity of gravel used, limits inter-granular contact and prevents the risk of blockages in highly reinforced situations.

Obtaining adequate paste fluidity when the paste has an increased concentration of fine particles requires the introduction of fluidizing chemical additives. Super-plasticizers, which act like deflocculating agents for the finest particles, are regularly made use of.

In order to limit the risk of the largest aggregates in suspension in the paste separating out, the paste viscosity must be adjusted to slow the phenomenon down, or even suppress it. Thickening agent additives are therefore used.

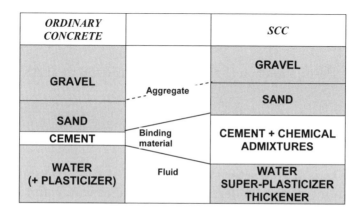

Figure 1.3. *Schematic comparison between compositions of SCCs and conventional vibrated concretes*

Note that the principles for formulations summarized above do not contradict the requirement for a compact grain size spectrum. The component concentrations in a SCC can, in effect, be determined in order to optimize the packing density of the granular mixture. The optimization method developed by de Larrard [LAR 00], can be used to this end.

SCC formulations must also include an objective in terms of target strength, denoted f_c, for the hardened mixture. Since SCC components are the same type as those for conventional concrete, it is logical to make use of the usual relations which link the target strength to the component concentrations.

The Bolomey equation [BOL 35]:

$$f_c = k_b R_c \left(\frac{C'}{E+V} - 0.5 \right) \qquad [1.1]$$

where R_c is the true class of the binder, k_b is a characteristic coefficient of the aggregates, C' is the mass of equivalent binder, E is the mass of water and V is the mass of water which occupies the volume of the entrapped air void. This equation is currently used for conventional concretes.

The Feret equation [FER 1892] generalized by [LAR 00]:

$$f_c = k_g R_c \left(\frac{V_c}{V_c + V_e + V_a} \right)^2 EMP^{-0.13} \qquad [1.2]$$

where k_g is a characteristic coefficient of the aggregates, V_c the volume of the binder, V_e the volume of water and V_a the volume of the entrapped air void. EMP is the maximum paste thickness (maximum distance between two large grains in the mixture) [LAR 00].

This type of equation is preferred for concretes which have fine aggregates such as sand concretes [ENP 94] and mortars. Nevertheless, in the case of SCCs, the Bolomey equation is still in current usage. The equivalent binder concepts can also be used to take into account the effect on strength of introducing mineral additives. The amount of entrapped air voids is an important parameter for SCCs. In a few cases, the introduction of an air-entraining agent in the formulation is foreseen. This development has a double role. On the one hand, creating air bubbles within the paste modifies its viscosity. On the other hand, the presence of air

bubbles can lead to a change in the size distribution of occlusions, which affect the conditions of grain arrangement.

Component	Mass quantities range (kg per cubic meter of concrete)	Volumetric quantities range (liter per cubic meter of concrete)
Powder (cement, etc.)	380-600	
Paste		300 – 380
Water	150/210	150/210
Gravel	750/1,000	210 – 360
Sand	Typically 48-55% of the total aggregates	
Water/binder		0.85/1.1

Table 1.1. *Range of typical mixture proportions for SCCs [EFN 05]*

1.2.2. *Specificity of SCC formulation*

As has been previously hinted at, formulating SCC involves increasing the proportion of fine particles in the mixture. Increasing the proportion of cement in the mixture is a satisfactory solution from a technical point of view, but often penalized, economically speaking, and detrimental to the environment (energy consumed and greenhouse gas emissions). Using additives is therefore often the preferred route. The introduction of chemical additives is generally required to ensure the fluidity of the mixture. Each component plays a different role with regard to the rheological behavior of the concrete, and some components interact with one another.

1.2.2.1. *Effects of chemical additives (super-plasticizer and thickening agents)*

Super-plasticizers work by dispersing particles of cement and mineral admixture. They effect this dispersion in two modes, the dominant mode depends on the nature of the basic polymer. An electrostatic equilibrium is achieved by neutralizing the positive electric charges which exist on the surfaces of cement grains, and sometimes mineral additives. A steric repulsion is induced by the presence of macromolecules coating the particles of cement and mineral admixture.

Certain polymers are branched (for example polyoxyethylene) which are not adsorbed and which persist in the water surrounding the particles. These molecules disturb the flow of fluid between the particles. Contact between particles is almost impossible and the film coating the particles modifies the lubrication between particles. The result of these two modes of action is a fluidizing of the mixture without water addition, or accompanied with a reduction in the water proportion required in the mixture.

Thickeners present a cohesive action in the same manner as the addition of fine particles in a paste. The AFGC recommendations require the use of these additives as soon as the water/binder ratio is too high and may induce segregation or bleeding [AFG 08]. By adding a thickener to a mixture, the mixture becomes more "sticky" which relates to an increase in the viscosity. In general, three principal action mechanisms are used to explain the increase in viscosity: adsorption, association and interlacing [KHA 98].

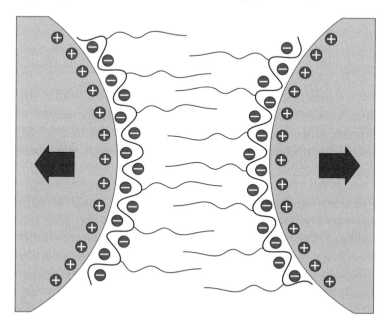

Figure 1.4. *Dispersion combining electrostatic and steric effects between two particles (cement and / or additive) [CAL 98]*

The first mechanism operates when the thickening agents have long chains (hydrophilic) which adsorb and fix molecules of water in the suspension, and extend throughout the mixture. From this an increase in the viscosity of the fluid arises, and therefore in that of the paste too. The second mechanism appears when adjacent thickener chains develop forces attractive to one another, thereby blocking the flow of water in the suspension. As a consequence, the suspension gels and flowing becomes more difficult.

Making this gel flow requires a shear stress level which is sufficient to break bonds. This is known as the static yield point. This mode of action therefore relates to a yield point value and an increase in the viscosity. Polymer interlacing generally occurs at a low shear rate and especially when the thickener concentration in the mixture is high. The chains interlace and tangle, causing the viscosity of the mixture to increase. Interlacing can be reduced at high shear rate by the alignment of polymer chains in the direction of shearing.

In a practical sense, the combination of the super-plasticizer's fluidizing action and the stabilizing action of the thickener show the fundamental characteristics of SCCs: fluidity and homogeneity. This combination is discussed in section 1.3.3.2.

However, the use of super-plasticizers and thickeners necessitates a test of their compatibility with the cement being used. This is carried out with laboratory tests [SCH 00, LAR 96]. As for the formulation of a conventional concrete, the cement used in SSC formulations does not require any particular specification. It must simply conform to the NF EN 197-1 standard, adapted to the target strength range and environmental class (NF EN 206-1).

The paste volume can be increased by adding a mineral additive. The mineral additive also contributes to avoiding an increase in the exothermic heat during the concrete's hardening phase if the concrete contains nothing but cement. Furthermore, an increase in the mixture proportion of Portland cement can induce increased shrinkage which may lead to cracking in the SCC.

At present in France, limestone filler tends to be the most widely used additive in SCC designs since it is not very costly and is available in large quantities, even though there is a reduction in the production of fly ash due to the progressive closures of coal-fired power stations.

1.2.2.2. *Significance of paste volume and gravel/sand ratio G/S*

An experimental study on this aspect [YAM 07], aiming to modify the volume proportion of paste (for a given paste formulation) in concrete, showed that three rheological domains exist which can be expressed as a function of the volume fraction of grains ϕ_{G+S} which represents the proportion of the mixture occupied by sand and gravel:

– for ϕ_{G+S} > 70%, concrete fluidity is low and concrete static yield point is high (> 1,000 Pa); grain interactions are likely to be dominated by direct friction contact, as is the case for conventional concrete;

– for 60% < ϕ_{G+S} < 70%, a transition regime is observed in which hydrodynamic effects and friction contact between grains act jointly and affect the behavior; the static yield point changes significantly;

– for ϕ_{G+S} < 60%, the yield points are low, in the order of tens of Pascals; *a priori*, hydrodynamic interactions dominate and direct contact is rare; since there is a large volume of paste, grain contacts are lubricated in the way that is desired for SCC.

Hence, the volume fraction of paste is commonly set between 35% and 40% in a SCC composition.

The size distribution of aggregates (sand and gravel) can interfere with the optimum paste volumes described above. In the case of conventional vibrated concretes, the mixture proportions of sand and gravel are calculated to yield an optimum compaction. This work can be carried out using grading curves according to the formulation methods developed by [DRE 70], or more judiciously with the aid of predictive calculation tools such as Bétonlabpro [LAR 99].

Studies on conventional concretes show that the grain size spectrum affects the strength of the hardened concrete. For a long time, a G/S ratio of around 2 was considered optimal for obtaining maximum compaction and therefore the highest strength possible. This was relevant, insofar as customary use was limited to rather coarse sand 0/4 mm and large gravel 4/20 mm.

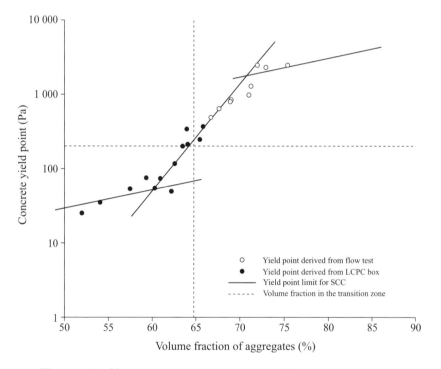

Figure 1.5. *Changes in a concrete's static yield point in response to changes in the solid volume fraction [YAM 07]*

Formulations of fluid concretes involve three, sometimes four, very fine grain size classes which reduce the G/S ratio to around 1.5 to achieve maximum compaction.

In the case of SCCs, a better concrete stability with regard to the segregation problem has been experimentally determined when the G/S ratio reduces to 1. In any case, values close to 1 are used in the majority of SCC formulations.

1.2.3. *Design methods for SCC*

A variety of SCC design methods exists. Globally, three broad classes of methodology are distinguished. The first is based on laboratory experiments. The second uses statistical data analysis, whilst the third approach is based on optimizing the paste volume around the aggregates.

1.2.3.1. *Experimentally determined formulations*

With regard to the research phase, formulations often seem to rest on the empiricism of the approach and a series of experiments may end up with, for example, a method which sets *a priori* a gravel/sand mass ratio of 1, a mixture proportion of 350 kg/m^3 cement, etc. This approach is hampered by a lack of reliability and is deficient as soon as one of the parameters varies slightly, for example as a result of absorbing water or changes in aggregate shapes.

A first step corresponds to synthesizing formulation principles from various published research works and adapting these principles to the manufacturing conditions (taking into account the characteristics of the materials used). The second stage comprises verification and validation using the research tests. This is indispensible for proving the robustness of the mixture.

In a more rational fashion, a second methodological class developed by the pioneers of SCC, for example Okumura *et al.* [OZA 89], enables the proposition of a relatively simple procedure based on the following principles: limited gravel concentration in the concrete (to ensure a low G/S), low water/binder ratio. The basic steps in creating mixtures are:

– setting the gravel volume to 50% of the solid volume in the mixture;

– setting the sand proportion to 40% of the mortar volume;

– water/cement volume ratio between 0.9 and 1.0 depending on the cement type; the corresponding mass ratio is very low, between 0.29 and 0.32;

– super-plasticizer and final water quantities, adjusted to ensure self-compacting ability.

The optimum amount of super-plasticizer can always be estimated from studies on mortar (cement + sand + water + additives).

1.2.3.2. *Formulations from statistical analysis*

This method is concerned with optimizing mixture proportions with the help of statistical analysis methods for the five fundamental formulation parameters:

– cement content C;

– mass ratio W/C;

– super-plasticizer concentration;

– thickener concentration;

– volume of large aggregates.

This approach was developed for example by Khayat *et al.* [YAH 01, YAH 03]. This method lends itself to an important experimental investigation. Laboratory tests (flow through a V funnel, yield point, viscosity, compressive strength at 7 and 28 days of age, etc.) which were intended to express the rheological and mechanical properties of fresh and hardened concrete, respectively, in terms of relevant parameters, were based on a wide range of variation in design parameters.

Analysis of the results data led to the construction of trend lines based on regression of experimental results. Linked with statistical models, the mixture proportions were optimized for a given rheology and strength which enabled a significant reduction in the number of laboratory tests.

1.2.3.3. *Formulation from packing density models*

These methods are certainly the most developed at the time of writing. Their origin is linked to the advent of high performance concretes, with their formulations intended to optimize the packing density. The models integrate a multi-scale analysis of the aggregate mixture proportions, from the largest to the finest particles.

In the case of SCCs, this approach can be used to provide a precise evaluation of the characteristics and quantity of paste to use for a given grain size spectrum. The paste quantity which fills up the voids between the aggregates follows directly from knowledge of how the aggregates are piled up; the space existing between the aggregates is inferred by an additional quantity of paste, called excess paste. The proportion of excess paste is calculated by fixing the ratio between the volume occupied by just the aggregates and the volume occupied by the aggregates covered in paste.

Some studies have shown that it is possible to create SCC by combining a paste which is simultaneously fluid and homogenous, optimized using the technique of mixture experiments with a granular skeleton for which the packing density and water requirement are known [BAR 05].

This approach is legitimate for G/S ratios between 0.8 and 1.3. For this kind of formulation, the packing density of aggregates can be determined using a compressible packing model [LAR 00], in order to take into account both the interactions of grains in the skeleton and also the effect of inner surfaces exerted by the container on the skeleton.

1.2.3.4. *Synthesis*

The formulation methods detailed above do not allow direct integration of the SCC fluidity criteria into the procedure. As a consequence, fluidity is in practice systematically adjusted or controlled with the aid of a series of tests in which the mixture proportions of additives and water are adjusted.

The idea of adjusting the chemical additives to concrete via tests carried out on pastes or mortars is very attractive since it rests on the principle of determining a relationship between the rheological behavior of the paste and that of mortar or concrete.

This scaling approach is physically relevant. However, it demands that test methods be available to enable estimations of intrinsic parameters of the materials tested. This problem is tackled in the following section and various results of rheological studies of pastes, mortars and SCCs are presented.

1.3. SCC rheology

1.3.1. *Fundamental concepts*

In France, "maniability" of concrete is defined as being the capability of the concrete to get into place under the force of its own weight with or without the aid of vibration. In North America, the workability of concrete is more widely discussed, defined as being a characteristic which determines the capability of a concrete or mortar to stay homogenous during mixing, casting, compacting and finishing.

Concrete workability is a generic term which refers to its aptitude for flowing, otherwise known as its rheology. Rheology, a word created by Bingham in 1929, is the science of flow dynamics, of deformations and more generally of the viscosity, elasticity and plasticity of materials under the influence of stresses.

In order to characterize the rheology of a paste, mortar or concrete, a rheometer is used. This apparatus is used to apply shearing to a sample, generally at a controlled shear rate, whilst measuring the induced shear stress τ.

The operation is automatically repeated at several shear rates. A diagram of shear stress against shear rate is obtained. This diagram is usually called the flow curve.

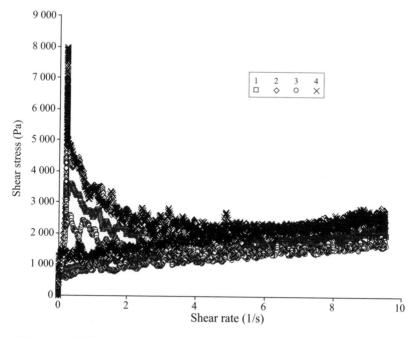

Figure 1.6. *Flow curve obtained for a conventional concrete (4 samples) tested with a Vane test rheometer [LAN 09]*

Figures 1.6 and 1.7 enable the flow curves of two types of concrete to be compared. The test involves increasing the shear rate, then decreasing it.

Each sample tested corresponds to a different batch. The uncertainty induced by the lack of reproducibility in concrete is evident.

However, the flow curve is characterized by a peak shear stress at the start of the test, in the case of conventional concrete. A similar peak is not observed in the case of SCC. When the shear rate is decreased, the linear nature of the flow curve is evident.

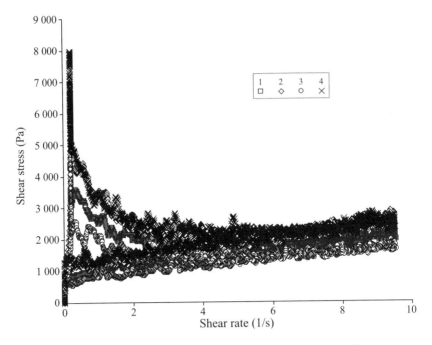

Figure 1.7. *Flow curve for SCC (4 samples) tested with a Vane test rheometer [LAN 09]*

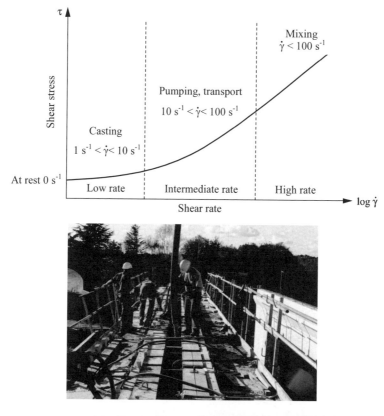

Figure 1.8. *Mixing, transport, pumping and pressurizing operations are all affected by concrete rheology. The shearing rate changes as a function of the procedure*

Converting the flow curve into characteristic and relevant parameters is useful. The most widely used parameters are (see Figure 1.9):

– the static yield point; this threshold corresponds to the shear stress necessary to induce the material to flow and can be modified by any structural features of the material (action of chemical admixtures, rest time before agitation, etc.); the reproducibility of this parameter is therefore difficult to evaluate;

– the dynamic yield point τ_0 is the shearing stress for a zero shear rate extrapolated from the linear section of the flow curve (characteristic of a destructured material);

– plastic viscosity η, the gradient of the linear section of the flow curve (characteristic of a destructured material.

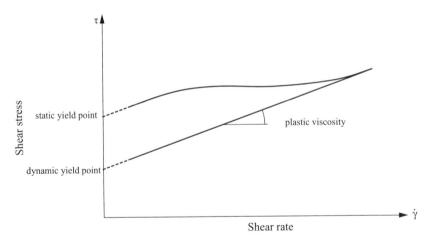

Figure 1.9. *Characteristic parameters calculated from the flow curve*

The dynamic yield point and plastic viscosity are the two parameters used in Bingham's linear behavior model:

$$\tau = \tau_0 + \eta \cdot \dot{\gamma} \qquad\qquad [1.3]$$

Equation [1.3] relates the shear rate $\dot\gamma$ (correlated to the rotational speed imposed on the shearing tool) to the shear stress τ deduced from the necessary torque at each rotational speed; η is the plastic viscosity while τ_0 is the dynamic yield point.

Other models can be fitted to the flow curve of the destructured material in order to take into account properties such as shear-thickening or shear-thinning behavior of the formulation being tested.

For certain concretes [LAR 98] or pastes [TOU 04] the best adapted models are non-linear. The Herschel-Bulkley or Casson models are preferred (see Table 1.2).

Equation from the model	Model
$\tau = \tau_0 + \eta_{pl}\dot\gamma$	Bingham [BIN 19]
$\tau = \tau_0 + \eta_{pl}\dot\gamma + c\dot\gamma^2$	Modified Bingham [YAH 01]
$\tau = \tau_0 + K\dot\gamma^n$	Herschel-Bulkley [HER 26]
$\tau = \tau_0 + \eta_\infty\dot\gamma + 2\left(\sqrt{\tau_0\eta_\infty}\right)\sqrt{\dot\gamma}$	Casson [CAS 59]
$\tau = \tau_0 + \eta_{pl}\dot\gamma e^{-\alpha\dot\gamma}$	De Kee [DEK 94]
$\tau = \tau_0 + 2\left(\sqrt{\tau_0\eta_\infty}\right)\sqrt{\dot\gamma e^{-\alpha\dot\gamma}}$	Yahia-Khayat [YAH 03]

Table 1.2. *Selected standard rheological models for concretes*

where:

– τ_0 is the dynamic yield point (Pa);

– η_{pl} is the plastic viscosity (Pa s);

– $\dot{\gamma}$ is the shear rate(s^{-1});

– c is the empirical constant;

– K is the consistency;

– n is the power index which represents the deviation from Newtonian behavior;

– α is the time-dependent parameter;

– η_∞ is the apparent viscosity at a very high shear rate.

1.3.2. *Rheological characteristics: methods and ranges of measured values*

To characterize a fluid concrete, the dynamic yield point and plastic viscosity are clearly the minimum pieces of information required, the static yield point of the structured material and time-dependent properties (thixotropy) are characteristics of complementary interest, with regard in particular to flow phases. Precise determination of these parameters is possible only with the aid of a concrete rheometer.

The cost of this equipment (which is very low in comparison with the cost of the quantities of concrete produced), its complexity when used on an industrial scale and perhaps the reduced number of available equipment types (BML CEMAGREF, BTRHEOM, IBB and two points test, Vane test, etc.) [LAR 96 and BAN 01] mean that their use is very low, almost for a very limited audience and restricted to research laboratories.

This aside, the principal problem that all these methods face is linked to calibrating the test device. The scattered results are a factor in considerably reducing operator confidence in the measurements obtained.

1.3.2.1. *Determining static and dynamic yield points*

The most common workability characterization tool used internationally is Abrams cone, which is used for the assessment of slump for concretes in general and for spread which is more specific for SCCs.

Slump corresponds to a loss in the height of the concrete sample after removal from the mold. This occurs through movement in the material induced by its own weight following extraction from the mold. This measurement is taken once the concrete becomes static again, i.e. when the shear rate induced by the movement is zero. [ROU 05] and [ROU 06] proposed a relationship between the sample height H after slump and the plasticity threshold τ_c of the material under test (specific gravity ρ) using the equation:

$$\tau_c = \rho.g.H/3^{0,5} \qquad\qquad [1.4]$$

In a conventional concrete prone to strong structuration, the shear rate induced by the flow is low and the effect of thixotropy remains noticed. In this case, the estimated plasticity threshold is very close to the static yield point (examples are given in [LAN 09]). Conventionally, the measured slump enables the classification of concretes as firm, plastic, very plastic and fluid without needing to calculate the static yield point.

In a concrete without any structuration effects, static and dynamic yield points are identical and estimated by equation [1.4].

In the case of SCCs, the sample, when freed from the mold, undergoes a significant flow. Its shape can change greatly and the parameter measured is therefore the sample spread. The time taken by a sample to reach a spread of 500 mm can be used as a complementary indicator. In this flow type, the shear rate at the sample core can be significant. The cessation of movement is therefore more characteristic of a destructured state of the material. As a consequence, the plasticity threshold calculated from the spread corresponds to the dynamic yield point. [ROU 05] and [ROU 06] proposed the following equation to calculate this parameter:

$$\tau_c = \frac{225\rho g \Omega^2}{128\pi R^5} \hspace{3cm} [1.5]$$

where Ω is the sample volume and R is the radius of the spread. This equation can be used for all types of reference volumes, from Abrams cone to a cylinder, for example. All that is required is the use of a sufficiently large sample.

Technically, it is easy to measure the spread, even on a construction site, without needing to resort to calculating the yield point.

For pastes and mortars, rheometers can be used and yield point measurements are easier to obtain. Using spread tests is also much easier.

1.3.2.2. Calculating apparent or plastic viscosity

For fluid concretes such as SCCs, the plastic viscosity must be measured. The time for the concrete to spread to 500 mm, which may be measured in a spread test, is an indicator of this viscosity. However, no relation between this time and the parameter value is suggested in the literature.

Work on SCCs has resulted in the emergence of numerous new methods of characterizing fresh concrete rheology which seek to take account of viscosity. A synthesis of these procedures has shown advantages, disadvantages and limits linked to shapes, costs and to the complexity of use of the more than 112 pieces of apparatus (U test, LCL box, Marsh cone, V-funnel, etc.) produced so far [FER 07].

The most direct indicators of the viscosity of a concentrated suspension are given by measuring the emptying time. With regard to concretes, the chosen shape for the test is the V-funnel. For pastes, the most well suited test is the cone test (Marsh or other cone). For these pieces of apparatus, various theoretical relations, which are rigorous to greater or lesser extents, between the rate measured, apparatus geometry and the fluid's rheological parameters have been suggested [ROU 05] and [ROU 06]. These relations enable the estimation of an equivalent Newtonian viscosity or parameters for a Bingham model (dynamic yield point and plastic viscosity).

At the time of writing, industrial practices of casting ready mix concretes on construction sites reveal that viscosity is never measured.

1.3.2.3. *Placing abilities and segregation*

In the presence of a significant reinforcement density, the flow of SCC must be checked for (granular) blockages which can lead to the formation of empty spaces or defects in covering the structure. In order to evaluate the passing ability, flow tests of the concrete in a highly reinforced environment are carried out (L-Box, Figure 1.10). After the flow stops, the levels reached by the concrete upstream and downstream of the reinforcements (H1 and H2, respectively) are measured. The filling rate is also determined from these values.

The segregation which can occur inside an SCC must be mastered. This phenomenon often appears after the SCC flow ceases or during flow at low shear rate. The large gravels, which cause a density gradient in relation to the surrounding paste, fall under the influence of gravity. The yield point and, to a lesser extent, the plastic viscosity of the concrete paste are the first order influences on this phenomenon. High values for these two properties are synonymous with low segregation. However, values which are too high result in concretes that are very firm and difficult to place, in contrast with the properties of SCC. A paste which develops structure rapidly provides a possible solution. In this case, however, maintaining SCC rheology over time is limited. A final possibility is to increase the paste density and/or to limit the density of the largest gravel in order to limit the extent of the phenomenon. [ROU 08] suggested the calculation of a function of the various parameters: the critical diameter dc above which there is a risk of segregation inside a viscoplastic fluid (dynamic yield point, τ_0):

$$d_c = \frac{18\tau_0}{\left|\rho_s - \rho_f\right| g} \qquad [1.6]$$

where ρ_s is the density of the gravel and ρ_f is the density of the fluid (the paste). In equation [1.6], the coefficient 18 can be adjusted as a function of the gravel shape and its roughness. Other authors suggested values from 10 to 25. As a consequence, for a grain size of 16 mm and a density difference of 600 kg/m^3, the minimum yield point for the paste must be in the order of 5 Pa (between 4 and 10 Pa).

The segregation risk, when considering finer particles, is more difficult to approach, since the definition of the carrier fluid is more debatable. Furthermore, this fluid probably does not show a yield point. Technically, the idea of segregation can be linked to the quantity of paste or fluid in excess which is present in the mixture. Removal by natural drainage of this excess fluid leads to gravel packing stabilization inside the SCC. The test used to calculate the proportion of this fluid which is susceptible to migration is the sieve stability test. A concrete sample is put on a 5 mm mesh sieve. The proportion of the fluid which passes through the sieve after two minutes is measured by weighing.

1.3.2.4. Construction site practices and recommended tests

In construction site conditions, simple and low-cost tools must be used to measure the various concrete behavior parameters. The French Association for Civil Engineering (AFGC) selected three tests which are the object of the recommendations. Globally, three characteristics are ensured:

– suitability for filling a formwork (filling ability), with the cone test (spread);

– suitability for filling a reinforced region (passing ability), determined using an L-Box;

– stability, evaluated using the sieve test (Figure 1.10).

Table 1.3 provides the limits for the characteristic values of an SCC. These limits are, for obvious reasons of effectiveness, not given in terms of characteristic values (yield point, viscosity, etc.) but in terms of values that are measured (spread, flow time, heights) by considering all the consistency classes defined in the EN 206-9 standard [AFN 10].

Standard method typical values	Class	Range or target values
Spread test with Abrams' cone	SF1 SF2 SF3	550 mm to 650 mm 660 mm to 750 mm 760 mm to 850 mm
500 mm flow time (spread test) t_{500}	VS1 VS2	≤ 2 s > 2 s
Mobility in an L-Box	PL1 PL2	H2/H1 $\geq$ 0.8 with 2 reinforcements H2/H1 $\geq$ 0.8 with 3 reinforcements
Sieve stability test	SR1 SR2	% laitance $\leq$ 20 % laitance $\leq$ 15

Table 1.3. *Acceptable values for SCC filling capabilities*

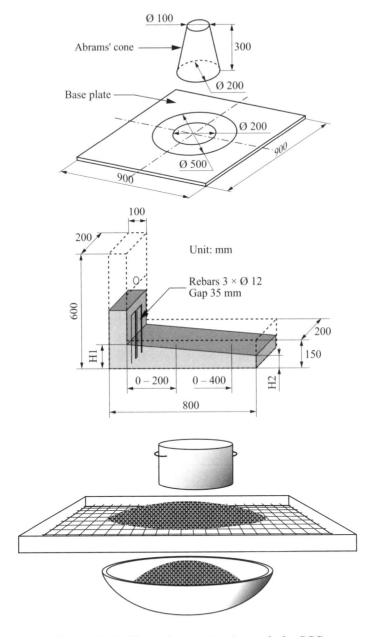

Figure 1.10. *Three characterization tools for SCCs (Abrams' cone, L-Box, sieve)*

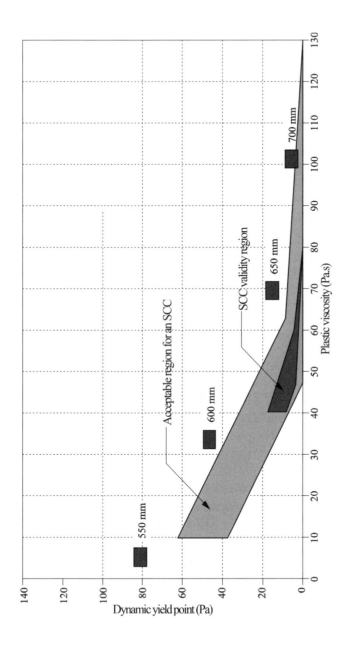

Figure 1.11. *Suitability diagram for an SCC defined by its plastic viscosity, dynamic yield point and spread diameter*

The plastic viscosity parameter is definitely the least well described in the AFCG recommendations. To this end, Wallevik [WAL 03] recommended that, in order to obtain the rheological properties of SCC, the concrete must have a combination of rheological parameters, the dynamic yield point and plastic viscosity, in a well-defined domain (Figure 1.11).

With a plastic viscosity less than or equal to 40 Pa.s, SCC must have a sufficiently high yield point, principally in order to reduce the risk of segregation.

On the other hand, if the self-compacting concrete is too viscous (plastic viscosity higher than 70 Pa.s), the yield point must be low in order to maintain an adequate filling capability. For a viscous concrete, the handling time is penalized. The minimum spread for obtaining SCC is also given as a function of viscosities.

However, it must be noted that the self-compacting character of a concrete is not simply an intrinsic quality. Appreciation of this quality must also take into account the structure to be made and as a function of the constraining conditions that the concrete will be in. Flow test results such as those obtained using an L-Box are very sensitive to the constraint type (steel mesh size, bar diameter, structure geometry, etc., in relation to the maximum particle diameter). These tests therefore risk being unrepresentative of real structures, which leads to wrongful rejection of some concretes and to the use of more expensive formulations.

By way of example, the choice of rheological characteristics can thus be made as a function of the application type (Table 1.4).

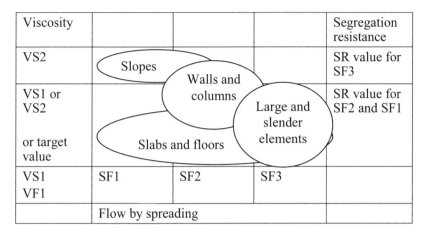

Viscosity				Segregation resistance
VS2	Slopes	Walls and columns		SR value for SF3
VS1 or VS2			Large and slender elements	SR value for SF2 and SF1
or target value	Slabs and floors			
VS1 VF1	SF1	SF2	SF3	
	Flow by spreading			

Table 1.4. *Envisioned applications as a function of SCC acceptability classes [WAL 03]*

1.3.3. *Rheology at different scales*

1.3.3.1. *From paste to concrete*

The paste, a mixture of cement, mineral additives, water and chemical additives, takes on a particular importance in SCC, which is only because it comprises a larger proportion of SCCs than of conventional vibrated concretes. It is therefore no great surprise to find a good number of studies in the literature which address the flow properties at the level of the paste by using the idea that, in the SCC context, the paste flow properties set those of the concrete.

For example, Saak *et al.* [SAA 01] showed that, from representing the balance of forces exerted on a particle suspended in the paste, optimization of the paste's rheological properties is needed in order to avoid particle segregation during and after casting the concrete (Figure 1.12).

Taking the essence of equation [1.6], Saak *et al.* proposed the identification of minimum and maximum values for these parameters which give rise to an acceptable SCC. The principle is illustrated in Figure 1.12.

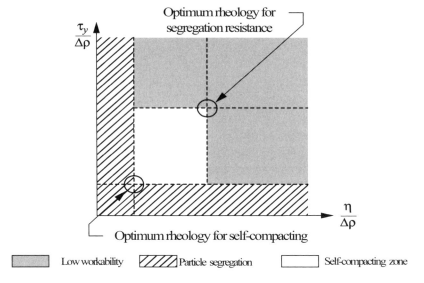

Figure 1.12. *Diagram to define the range for SCC yield point and plastic viscosity. Δρ is the difference in density between the gravel particles and paste [SAA 01]*

Since the rheological characterization of the paste is easier to determine than that of the concrete, it is useful to establish how the rheological properties of the paste can be used to determine those of the concrete.

The most commonly used approach involves viscosity models, or models with multi-scale suspensions, which allow the viscosity η of a mixture to be estimated from the viscosity of the suspending fluid η_f and the volumetric fraction of particles in suspension (ϕ). Some examples of viscosity models are shown in Table 1.5. The volumetric fraction of particles in SCC is generally between 50% and 65% (shown in Figure 1.5). Einstein's low concentration suspension model is therefore not applicable. Considering a maximum particle volumetric proportion (ϕ_m) of 85%, and an intrinsic viscosity of 2.5, a range of plastic viscosity for SCC from 40 to 80 Pa.s can be envisioned (Figure 1.11) which corresponds to a range of plastic viscosity for the paste between 1.84 and 12 Pa.s, if the Krieger-Dougherty model is used.

Model	Equation
Einstein [EIN 06]	$\eta = \eta_f (1 + 2.5\phi)$
Krieger-Dougherty [KRI 59]	$\eta = \eta_f \left(1 - \dfrac{\varphi}{\varphi_m} \right)^{-[\eta]\varphi_m}$ where [η] is the intrinsic viscosity
YODEL [FLA 06]	$\tau_c = m_1 \left(\dfrac{\phi^3}{\phi_m(\phi_m - \phi)} \right)$ where m_1 is a constant
Changing the scale on a suspension [CHA 08 and MAH 08]	$\tau_c(\phi) = \tau_{cf} \left(\dfrac{\sqrt{1-\phi}}{\left(1 - \dfrac{\phi}{\phi_m} \right)^{2,5\phi_m}} \right)$ $\eta = \eta_f \left(1 - \dfrac{\phi}{\phi_m} \right)^{-2,5\phi_m}$

Table 1.5. *Viscosity and yield point models*

However, it must be noted that the viscosity models presented in Table 1.5 are theoretically usable with a viscous Newtonian fluid. SCCs are viscoplastic fluids. With regard to the yield point, some authors have suggested the use of a Krieger-Dougherty model written in terms of the yield point. The YODEL model [FLA 06] is an efficient alternative.

Work by [CHA 08] and [MAH 08] contributed some new elements in the case of pastes which can be modeled as viscoplastic. In the case of a Bingham fluid, the suspension yield point τ_c is linked to the paste yield point τ_{cf} while the plastic viscosity of the suspension η is linked to that of the paste η_f. These theoretical models have yet to be validated, since they correspond to perfectly spherical occlusions.

According to this approach, when ϕ varies from 50 to 65%, ϕ_m is 85% and the SCC's yield point is between 5 and 70 Pa (limit of non-segregation of large particles and upper limit in Figure 1.5), the yield point of the paste must be between 0.4 and 15 Pa.

1.3.3.2. *Understanding the role of constituents at the paste level*

Optimizing a SCC paste relies on knowledge of the role played by its components in its flow properties, all whilst taking into account the interactions which may exist between the components. In effect, it has been shown that super-plasticizers and thickening agents have a detrimental effect on the static yield point and viscosity [SVE 03, DAL 06]. As an example, Figure 1.13 shows that, for fixed proportions of thickening agent (or super-plasticizer), the yield point and the viscosity decrease (or increase) whilst the proportions of the super-plasticizer (or thickening agent) increase.

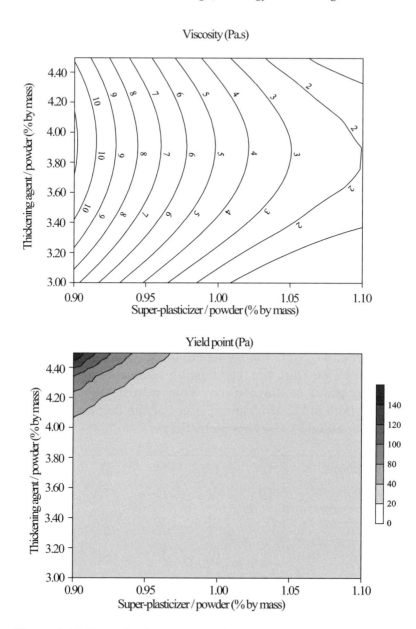

Figure 1.13. *Example of antagonist effects between the super-plasticizer and the thickening agent on the yield point (bottom) and the apparent viscosity (gradient 18 s⁻¹) (top) [ELB 09]*

In practice, this means that an optimum combination of proportions for these two additives must be found when using them simultaneously (particularly in horizontal structures). To this end, the technique which uses mixture experiments [COR 90, MAT 00a, MAT 00b] is a great help in the laboratory since it allows variation in the proportions of components, and implicit proportions of other components, to be taken into account in the context of experiments in which all mixtures have measurable properties, bearing in mind their relational constraints. This technique also allows, when used with statistical analysis, the influence of constituents on the flow properties of the paste to be ranked in order of significance (Table 1.6).

The yield point, which reflects a greater or lesser structuration that impedes the flow, is affected by the introduction of super-plasticizing (SP) additives and thickening agents (TA), The mechanism of this antagonistic action means that the proportion combinations must be determined so that the yield point remains low, all whilst reducing grain sedimentation. This returns to a stabilized state thanks to the thickening agent of the dispersed particle system enabled by the super-plasticizer.

Yield point	SP	TA	C	W	F
Spread	SP	W	C	TA	F
Viscosity ($4s^{-1}$)	C	SP	W	TA	F
Viscosity ($18s^{-1}$)	C	W	F	SP, TA	
+ ← Impact → - SP=super-plasticizer, TA=thickening agent, C=cement, W=water, F=limestone filler					

Table 1.6. *Hierarchical influence of constituents on the flow properties of the paste [ELB 09]*

Good spreading, which results from a low yield point, is naturally determined by the SP proportion which is the most influential factor. The proportions of water (W) and cement (C) are secondary since the size of the spread is affected by the W/C ratio.

Essentially, the viscosities reflect the effects linked to inter-particle friction and it is to be expected that, with regard to these parameters, the cement takes first place since it is the principal component of the suspension. During the flow, the SP effect is less sensitive, seemingly due to the alignment of its chains in the direction of shearing, which reduces inter-particle separation and increases friction. The second proportion parameter which affects the viscosity is that of water, followed by the proportion of fine particles in the mixture.

The thickening agent does not seem to play an important role with regard to variations in spread or viscosity. More logically, the TA acts more on the stability of the mixture at rest.

This parametric study is a simple example. Modifying one of the constituents (a change in the nature of the cement, fine particles or chemical additive) will affect the results. Optimizing the paste formulation is a difficult step to learn without at least a little work in the laboratory.

1.3.4. *Evolution in rheology during casting – thixotropy*

In practice, a concrete is thixotropic, which means that it behaves like a reversible fluid with slow restructuring, following a rapid destructuring. More precisely, it develops structure, when at rest, in a characteristic time of around 100 seconds and destructures in a characteristic time of tens of seconds at a shear rate in the order of some s^{-1}. All concretes are thixotropic, and SCCs are even more so.

From a rheological perspective, the thixotropy phenomenon is observed when the material is at rest. The more rapidly the material solidifies, the more thixotropic it is. This is the case even for SCCs with low water content, whilst the content of fine particles is high. The action of super-plasticizers, associated with an adapted granular composition, allows a high fluidity when the material is destructured, with a very low yield point in this state (<100 Pa) which increases rapidly when the material is at rest. Observations show that the rate of increase in the yield point is constant [AMZ 08].

From a practical perspective, a strong thixotropy has the advantage of making the pressure in a formwork decrease rapidly, as will be seen later. However, in the case of casting floors, a rapid structuring gives rise to phenomena similar to a cold joint when a layer of fluid concrete covers a still layer already in place. The critical time for these phenomena to appear is in the order of 30 min when the structuration rate at rest is 0.3-0.5 Pa/s.

1.4. Industrial practices

1.4.1. *Determining rheology during mixing and transport*

Mixing is the first stage in the concrete fabrication process (Figure 1.14). The structural homogeneity of the concrete depends on the mixing quality. In this regard, work from the Laboratoire Central des Ponts et Chaussées (LCPC) has shown the influence of the mixing parameter on the concrete's homogeneity as it leaves the mix truck and on the rheological characterization in the laboratory mixer or the mixing plant [CAZ 09]. Hence it has been shown that physical measurements can be determined from the power curve recorded during mixing, such as the yield point and plastic viscosity of the concrete [CHO 04] or the apparent characteristic measurements according to the principles of the two points test [TAT 73].

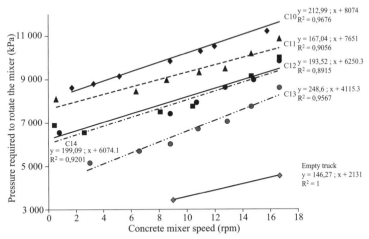

Figure 1.14. *Mixing concrete 700 BC. Evolution towards a standard mix truck from which a rheogram can be obtained, comparable to those obtained using a laboratory rheometer*

In the case of SCC, low yield points lead to torque values at the mixer axis and therefore values of consumed electrical power that are fairly different to those obtained for conventional concrete. The use of these measures for characterizing SCC can therefore still be difficult; the effect of a small variation in composition can result in significant variation in the measurement.

Analysis of the concrete fabrication process shows that the transport phase, particularly just before casting, is definitely the most suitable for measuring the rheological properties of the concrete (Figure 1.15). This is also the usual time chosen to carry out an Abrams cone test. Moreover, this enables verification that the transport time has not affected the desired rheology with the selected mixture proportions.

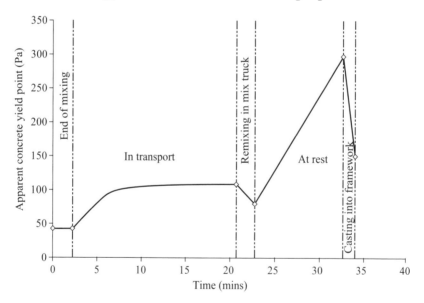

Figure 1.15. *Apparent yield point kinetics of SCC throughout the fabrication process (mixing, transport, filling of the formwork)*

For these reasons, the possibility of carrying out rheological measurements via the use of available and measurable data in the truck at this critical moment has been evaluated [AMZ 06, AMZ 07]. A variable rotation speed of the mixing tank on the truck can be set (1 to 18 rpm). The power required to rotate the loaded mixing tank at each speed is measured (Figure 1.14).

From this data, a rheogram of the tank rotation speed with mixing can be drawn which is similar to that obtained with a rheometer. A good correlation between the parameter obtained from the mixing truck equivalent to the yield point and the parameter from a concrete rheometer is observed. Some disagreements in the results for the plastic viscosity, an essential parameter for characterizing an SCC, are still observed.

This rheometric approach is still quite technically easy to carry out without requiring the use of further apparatus (for example rheometers). Taking a measurement during mixing with an immersed probe is a viable alternative [CAZ 07]. However, this general approach do not allow the identification of intrinsic concrete parameters and the corresponding relations between these general parameters and the intrinsic rheological parameters are yet to be determined.

1.4.2. *Pumping*

1.4.2.1. *Definition*

Pumping is a frequently used process on construction sites to transport concrete from the mixer or mix truck to where the casting will take place. The distances to traverse can therefore reach several hundred meters high, and up to approximately 2 km horizontally. By way of illustration, on a high level construction site in Dubai, concrete was pumped up to an extremely high level of just over 500 m. Concrete is transported at pressures of over 500 bars. By using a pump, concrete is pushed along in steel or elastomer pipes according to the pressure used.

In order to use pumping, the concrete must be considered to be "pumpable". The pumpability is usually defined as the concrete's mobility and stability under pressure. It is concerned with the capability of a fresh concrete to move under the influence of pressure in a confined space, all the while retaining its initial properties [GRA 62].

1.4.2.2. *Influence of composition and rheology on pumping*

All of a concrete's composition factors affect its suitability for pumping. Concrete has an initially void volume of 7-10% when leaving the mixer (before vibration). For current concretes, two methods are available for improving pumpability: either reducing the void volume or increasing the paste volume.

The customary formulation rules are:

– aim for an overall granularity (fine and coarse gravel combined) characterized by a well spread curve; in general, by aiming for a spread out granularity, a maximum compactness is obtained;

– keep a minimum proportion of fine particles in the mixture; in some cases, some authors have suggested the combined use of a grading curve and a minimum content of fines;

– consider a maximum water content beyond which the segregation risk is increased.

In summary, optimized grading results in a minimum content of voids in the granular mixture, which allows the necessary volume of paste to be limited. Note that ideal granularities are, in general, rich in fine particles.

The characteristics of SCC formulations (significant paste volume, low W/C, addition of a thickener to improve stability) produce a material which is well suited to pumped transport, as long as the gravel selection is in keeping with an optimum shape factor. In order to optimize granular compactness, the size distribution should be adjusted to avoid abrupt jumps in the grading curve. The size and shape of the particles and their roughness must therefore be taken into account since these factors greatly influence the compactness of the mixture [KEM 69].

Concrete's pumpability is therefore fundamentally dependent on its rheology upon exit from the mixer. Furthermore, the stability of the rheology must be ensured throughout the duration of the handling, knowing that, in a real-world situation on a construction site, technical and mechanical constraints linked to the choice of material are not possible to master. As a consequence, pumpability is a characteristic which depends uniquely on the properties of the concrete when fresh, and stays independent of equipment and pumping conditions. The factors which determine pumpability can therefore be confirmed to remain linked to the concrete composition [CHO 04].

1.4.2.3. *Modeling pumping*

Assuming a lubricating layer with Bingham behavior and a central core of concrete which also has Bingham behavior (Figure 1.16), the overall flow Q$_{tot}$ (equation [1.7]) is the sum of the shear flow (described by the Buckingam-Reiner equation) and the sliding flow (Qs):

$$Q_{tot} = \frac{\pi R^4}{8\eta} \frac{dp}{dx} \left[1 - \frac{4}{3} \cdot \frac{2\tau_0 dx}{R dp} + \frac{1}{3} \cdot \left(\frac{2\tau_0 dx}{R dp} \right)^4 \right] \cdot k_r \cdot 3600 + Q_s$$

[1.7]

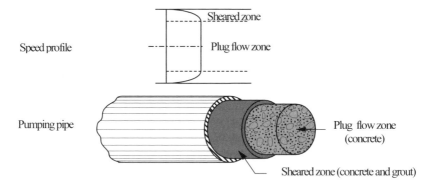

Figure 1.16. *Speed profile in a pumping pipe*

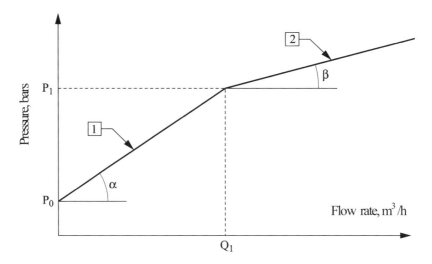

Figure 1.17. *Diagram of pressure with flow during concrete pumping*

Kaplan used this hypothesis to establish a model [KAP 00]. He neglected the fourth power term in the part of the equation for the shear flow and used parameters determined by a tribometer (yield point at the interface and plastic viscosity at the interface) to define the sliding flow.

By calculating the shear rate at which supplementary flow is initiated in the central section (in the concrete) and by reorganizing the terms of the equation, Kaplan expressed the pumping pressure P as a function of the total resultant flow (Figure 1.17). The break in Figure 1.17 corresponds to the appearance of the shear flow created by pressure losses which result in a loss of sliding flow.

The first slope is described by equation [1.8]:

$$P = \frac{2L}{R}\left[\tau_{0i} + \frac{Q}{3600\pi R^2 k_r}\eta_i\right] \qquad [1.8]$$

where:

- P is the pressure (Pa);

- L is the pipe length (m);

- R is the pipe radius (m);

- Q is the average flow rate (m³/h);

- k_r is the filling coefficient;

- τ_{0i} is the yield point at the interface (Pa);

- η_i is the plastic viscosity at the interface (Pas/m).

This equation is valid when $P \geq \dfrac{2L}{R}\tau_{0i}$.

In phase 2, equation [1.9] takes over:

$$P = \frac{2L}{R}\left[\tau_{0i} + \frac{\dfrac{Q}{3600\pi R^2 k_r} - \dfrac{R}{4\eta}\tau_{0i} + \dfrac{R}{3\eta}\tau_{0i}}{1 + \dfrac{R}{4\eta}\eta_i}\eta_i\right] \qquad [1.9]$$

In this phase, shearing is higher than the concrete's yield point, and the concrete also flows. From then on, the rheological properties of the concrete as well as the properties at the interface must be taken into account. This point is definitely critical whenever it concerns an SCC; this may be the main reason why a generalized model has not yet been established for SCCs.

1.5. Forces exerted by SCCs on formworks

1.5.1. *Important parameters*

The increase in the forces exerted by fresh concrete on the inner surfaces of the formworks is a critical issue which has been the subject of numerous investigations reported in recent literature. In the case of SCCs, which can potentially be cast at great height, this phenomenon must be mastered since it affects the design of formwork equipment. Many parameters influence the forces exerted by concrete on a vertical inner surface. Despite this, many authors agree on the subject of the effects of these parameters on pressure. Investigations carried out by [ROD 52], [GAR 79], [GAR 80], [BIL 06], [AND 04] and [TCH 08] show that lateral pressure and its magnitude depend on:

– the concreting height;

– the casting speed;

– the concrete and ambient temperatures;

– the presence, or absence, of vibration;

– the concrete design;

– the concrete's consistency;

– the size and shape of aggregates and the type of cement;

– the size and shape of the formwork.

In order of importance, the dimensions of the formwork's inner surfaces (the height of the formwork, essentially), the concrete specific gravity, the casting speed and vibration are the principal parameters.

The most recent studies have clearly shown the importance of the workability parameter or more exactly of the yield point and its evolution with time. Hence it has been noticed that the inner surfaces of the formworks cause tangential frictional stresses which can reach the yield stress of the concrete in use.

These forces oppose the effects of gravity, thereby reducing the local pressure and also the lateral pressure of the concrete against the formwork. Furthermore, it has been shown that the evolution in pressure exactly follows the evolution of the yield point during structuration of the concrete (and during setting).

This result is important, insofar as it is henceforth possible to regulate the casting speed with the structuration speed of the concrete in use. Knowledge of an SCC's rheology thus contributes to making the practice of concreting at great heights safe.

1.5.2. *Changes in pressure against a formwork*

In order to illustrate the evolution of forces to which a formwork is submitted, the results of an experiment are presented which shows the lateral and vertical pressures in a formwork. The experimental mechanism (Figure 1.18) uses a formwork column made of a PVC tube with an interior diameter of 100 mm, thickness 5.3 mm and height 130 cm, for casting the material. The column is capable of withstanding pressures up to 10 bars with a waterproofing system at its base.

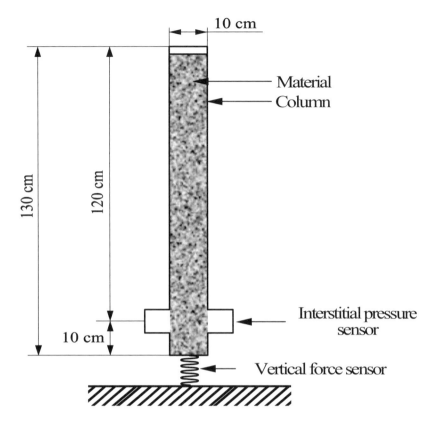

Figure 1.18. *Laboratory apparatus for measuring the different pressure components exerted by a concrete on its formwork*

The column is equipped with two interstitial pressure gauges [AMZ 06] situated at a depth of 120 cm. A vertical pressure gauge, with a load of 0.5 kN and an accuracy of ± 0.2%, coupled to a sliding system (piston) is also put at the bottom of the formwork at a depth of 130 cm.

The results are presented in Figure 1.19. Three phases are generally observed, as described below.

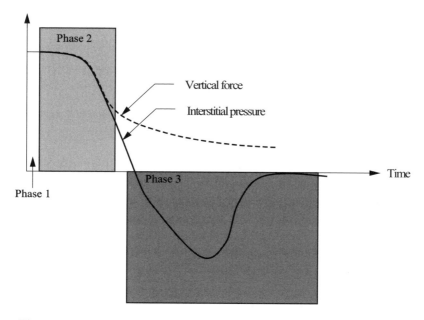

Figure 1.19. *Diagram of the change in interstitial pressure and vertical pressure as a function of time*

Phase 1. After filling the formwork with the material and vibrating it (Figure 1.18) an increase in pressure is observed with a peak that corresponds theoretically to the maximum hydrostatic pressure:

$$P_{th} = \rho.g.h \qquad\qquad [1.10]$$

and to a vertical load:

$$P_v = \rho.g.S.h \qquad\qquad [1.11]$$

where ρ is the specific gravity of the mixture in kg/m^3, h is the height in m of the material that has been cast into the formwork, g is the force of gravity in m/s^2 and S is the cross-sectional area of the column in m². As a result of contact effects between the material and surfaces, these theoretical values are not achieved. The differences relative to the theoretical values are a function of the friction at the interface. Hence determining the frictional forces is essential if the maximum pressure of material in a formwork is to be predicted.

Phase 2. The material is now at rest, a fall in interstitial pressure is observed for the three types of mixture. A fall in the vertical load is also observed, which reflects an increase in the load bearing by the inner surfaces of the formwork column through the mobilization of tangential stresses. The similarity between the changing kinetics of the relative lateral pressure and the relative vertical force is striking, regardless of the nature of the mixture tested. The simultaneous nature of the changes is observed up to a level of 50% of the maximum pressure. After this, a divergence is observed in the vertical pressure since the fall in interstitial pressure accelerates. The point of divergence is clearly a characteristic change in the behavior of the material under test. Structuration effects give rise to changes in the interstitial pressure and the vertical load in the formwork. These effects do not act in the same way on interstitial pressure and vertical load. The vertical force tends to stabilize. In contrast, mixtures can be distinguished in this phase by the speed of the fall in interstitial pressure, which is more rapid when the W/C ratio is low.

Phase 3. One of the distinguishing features in the structuration of cement-based materials is, without doubt, the appearance of a low pressure phase which results from the suction of water trapped in a capillary network formed by the cement particles during the hydration process. The moment at which this depressurization appears is closely linked to the W/C ratio of the cement paste.

In comparison with conventional concretes, the research by [TCH 08] shows that the addition of chemical admixture and filler has a slowing effect on the pressure kinetics. The granular effect is highly significant in the depressurizing phase since it considerably reduces the magnitude of negative pressures. This observation is a good reflection of the granular effect on the level of shrinkage strain variations which arise from the hydration of the cement paste. The aggregates significantly reduce the shrinkage stresses and limit their level in the case of concretes.

The presence of steel reinforcement also has an effect in the sense that a marked reduction in concrete pressure has been shown in this case. This aspect ought to be taken into account in regulations.

With regard to SCCs, a fairly high structuration time, relative to that of a conventional concrete, can lead to a limited reduction in pressure with slowed-down kinetics.

1.5.3. *Adapting the casting conditions*

Since the lateral pressure, initially equal to hydrostatic pressure, decreases after casting, it is possible to predict that the rate at which the pressure drops evolves in accordance with the thixotropic behavior of the concrete as described in section 1.3.4.

The CEBTP [CEB 99] measured the pressure exerted by SCC with a W/C ratio of 0.46 with a spread diameter of 700 mm on a surface of 12 m in height, 2 m long and 0.34 m thick. The flow rate was 25 m/h for a pumped concrete and 10 m/h for a poured concrete. The maximum wall pressure obtained was reduced by 70% and 65% of the hydrostatic pressure in each case respectively according to whether the casting was carried out by pumping or by injection at the bottom of the formwork. These results show a more significant wall pressure when the concrete is injected compared to when it is poured. This observation was confirmed by Leeman and Cuthlenen [LEE 03].

In this regard, Ovarlez and Roussel [OVA 06] proposed a model presented below, which enables the evolution of the lateral stress exerted by a self-compacting concrete on the surface of a formwork during and after casting to be described. A comparison of predictions from the model with measurements of the fall in lateral pressure at the end of casting associated with measurements of changes in the material's yield point lead to the following conclusions:

– lateral pressure is equal to the hydrostatic pressure when the casting speed is high or when concrete is cast by injection;

– lateral pressure does not go as high as the hydrostatic pressure when the casting is by pouring.

1.5.4. *Modeling pressure*

Using the relation proposed in equation [1.12] to characterize the time evolution of the static yield point induced by structuration effects, and through setting the start time of the experiments as the moment at which contact is made with water, the following equation is obtained:

$$\tau_0(t) = A_{thix} t \qquad\qquad [1.12]$$

Assuming that the casting speed, V, is constant, the concreting height is given as a function of time by the expression H = Vt. Three formwork types can be distinguished: a rectangular formwork of size "e", a circular column of radius "r" and finally the most general case of a formwork which contains reinforcements [PER 09].

The shear stresses inside the concrete vary with the height of the concreting and must be considered when calculating the lateral stress at the end of the formwork given by the expressions in Table 1.7.

Formwork cross-section shape	Equation for the evolution of lateral pressure
Rectangular cross-section formwork	$P = \left(\rho g H - \dfrac{A_{thix} H^2}{eV} \right)$
Circular cross-section formwork	$P = \left(\rho g H - \dfrac{A_{thix} H^2}{rV} \right)$
Circular cross-section formwork containing a reinforcement with diameter ϕ_b [PER 08]	$P = \left(\rho g H - \dfrac{\varphi_b + 2S_b}{(r - S_b)\varphi_b} \dfrac{A_{thix} H^2}{V} \right)$

P: Lateral pressure of the concrete in use; ρ specific gravity of the mixture in kg/m³; H height of the material cast in the formwork in m; g acceleration due to gravity in m/s²; e width under consideration; r radius of the formwork; V casting rate, ϕ_b diameter of reinforcement; S_b average rate of reinforcements per meter length of formwork

Table 1.7. *Pressure models affected by formwork conditions*

The relationships turn out to be capable of explaining the phenomena observed during experiments. They predict, in effect, that the lateral stress in the bottom formwork cross-section is limited as a function of H. They explain two regimes: in the first regime the influence of the casting speed on the lateral stress is very high, the second regime is at rest and is characterized by the influence of an increase in the yield point at rest.

The structuration rate is A_{thix}, taking into account this increase, which ranges from 0.1 Pa/s to 2 Pa/s and could be measured by means of rheological tests. The roughness of the formwork must also be estimated since changes in the yield point at the surface depend strongly on the state of the surface.

As was described earlier, the structuration rate depends on the mixture (and probably on the temperature) and introduces a factor from 1 to 20 times the relative pressure. If we consider that the limit, in practice, linked to the casting speed is between 1 m/h and 20 m/h, another factor of 1 to 20 is also introduced. It therefore appears that the structuration rate and the casting speed are the dominant parameters in this phenomenon.

Identification of the optimum formwork duration is an obvious interest in industry in order to optimize the formwork rotation. Today, although there are viable tools such as the thermal technique of maturometry developed by LCPC or acoustic auscultation, their use does not enable a good characterization of the evolution of the concrete's rheological properties during casting or hardening. A certain parallel can be proposed between the structuration model of fluid concrete at rest and of concrete during setting and hardening.

1.6. Bibliography

[AFG 08] AFGC, Recommandations pour l'emploi des bétons auto-plaçants, AFGC, January 2008.

[AFN 10] NF – EN 206-9, Béton, règles complémentaires pour le béton auto-plaçant, AFNOR, June 2010.

[AMZ 06a] AMZIANE S., "Setting time determination of cementitious materials based on measurements of the hydraulic pressure variations", *Cement and Concrete Research*, vol. 36, no. 2, p. 295-304, 2006.

[AMZ 06b] AMZIANE S., FERRARIS C.F. "Feasibility of using a concrete mixing as a rheometer", *NISTIR 7333*, 2006.

[AMZ 07] AMZIANE S., FERRARIS C. F., "Cementitious paste setting using rheological and pressure measurements", *ACI Materials Journal*, vol. 104, no. 2, p. 137-145, 2007.

[AMZ 08] AMZIANE S., PERROT A., LECOMPTE T., "A novel settling and structural build-up measurement method", *Measurement Science Technology*, vol. 19, no. 10, 105702 (8pp), 2008.

[AND 04] ANDRIAMANANTSILAVO N.R., AMZIANE S., "Maturation of fresh Cement Paste within 1-to-10-m-large-formworks", *Cement and Concrete Research*, vol. 34, no. 11, p. 2141-2152, 2004.

[ASS 03a] ASSAD J., KHAYAT K.H., MESBAH H., "Variation of formwork pressure with thixotropiy of self-consolidating concrete", *ACI Materials Journal*, vol. 100, no. 1, p. 29-37, 2003.

[ASS 03b] ASSAD J., KHAYAT K.H., MESBAH H., "Assessment of thixotropy of flowable and self-consolidating concrete", *ACI Matérials Journal*, vol. 100, no. 2, p. 99-107, 2003.

[BAN 01] BANFILL P.F.G., BEAUPRÉ D., CHAPDELAINE F., DE LARRARD F., DOMONE P., NACHBAUR L., SEDRAN T., WALLEVIK O., WALLEVIK J.E., "Comparison of concrete rheometers: International tests at LCPC Nantes, 2000" in C.F. FERRARIS, L.E. BROWER (eds), *NISTIR 6819*, 2001.

[BAR 97] BARON J., OLLIVIER J.P., *Les bétons. Bases et données pour leur formulation*, Eyrolles, Paris, 1997.

[BIN 19] BINGHAM E.C., GREEN H., "Paint, a plastic material and not a viscous liquid; the measurement of its mobility and yield value", *Proceedings of the American Society of Testing Materials, II*, 19, p. 640-664, 1919.

[BOL 35] BOLOMEY J., "Granulation et prévision de la résistance probable des bétons", *Travaux*, vol. 19, no. 30, p. 228-232, 1935.

[CAL 98] Calcia infos, Spécial adjuvants, no. 16, 1998.

[CAS 59] CASSON N.A., "A flow equation for pigment–oil suspensions of the printing ink type", in C.C. MILLS (ed.), *Rheology of Disperse Systems*, Pergamon Press, New York, 1959.

[CAS 05] CASSAR L. (Italcementi group), *Nouveaux matériaux bétons*, Ecole thématique CNRS-ATILH, Paris, 2005.

[CAZ 07] CAZACLIU B., "Mesures en continu dans le malaxeur: améliorer la régularité de la production du BAP", *Béton(s)*, no. 9, p. 79-80, March–April 2007.

[CEB 99] The validated technique of self compating concrete, Report no. B242-9-054, Centre Expérimental du Bâtiment et des Travaux Publics (CEBTP), 1999.

[CHA 08] CHATEAU X., OVARLEZ G., TRUNG K.L., "Homogenization approach to the behavior of suspensions of noncolloidal particles in yield stress fluids", *Journal of Rheology*, vol. 52, no. 2, p. 489-506, 2008.

[CHO 04] CHOPIN D., DE LARRARD F., CAZACLIU B., "Why do HPC and SCC require a longer mixing time?", *Cement and Concrete Research*, vol. 34, no. 12, p. 2237-2243, 2004.

[COR 90] CORNELL J.A., "How to run mixture experiments for product quality", *Milwaukee: American Society for Quality Control*, 1990.

[DAL 06] D'ALOIA SCHWARTZENTRUBER L., LE ROY R., CORDIN J., "Rheological behaviour of fresh cement pastes formulated from a self compacting concrete", *Cement and Concrete Research*, vol. 36, no. 7, p. 1203-1213, 2006.

[DEK 94] DE KEE D., CHAN MAN FONG C.F., "Rheological properties of structured fluids", *Polymer Engineering and Science*, vol. 34, no. 5, p. 438-445, 1994.

[DRE 70] DREUX G., *Guide pratique du béton*. Collection de l'ITBTP, Paris, 1970.

[EFN 05] EFNARC, The European Guidelines for Self-Compacting Concrete Specification, "Production and Use", May 2005

[EIN 06] EINSTEIN A., "Über die von der molekularkinetischen Theorie der Wärme geforderte Bewegung von in ruhenden Flüssigkeiten suspendierten Teilchen", *Ann. Phys.*, vol. 17, p. 549-560, 1906.

[ELB 05] EL BARRAK M., Contribution à l'étude de l'aptitude à l'écoulement des bétons auto-plaçant (BAP) à l'état frais, Thesis, Paul Sabatier University, Toulouse, 2005.

[ELB 09] EL BARRAK M., MOURET M., BASCOUL A., "Self-compacting concrete paste constituents: Hierarchical classification of their influence on flow properties of the paste", *Cement Concrete and Composites*, vol. 31, no. 1, p. 12-21, 2009.

[FER 92] FERET R., "Sur la compacité des mortiers hydrauliques", *Annales des ponts et chaussées*, série 7, vol. 4, p. 5-164, 1892.

[FER 06] FERRARIS CHIARA F., KOEHLER E., AMZIANE S. *et al.*, Report on Measurements of Workability and Rheology of Fresh Concrete, ACI 238.1R-08, ACI, 2006.

[FLA 06] FLATT R., BOWEN P., "Yodel: a yield stress model for suspensions", *Journal of American Ceramics Society*, vol. 89, p. 1244-1256, 2006.

[FRA 07] FranceBTP.com, "Le BAP: où en est-on en 2007?", *Magazine BTP Matériaux*, December 2007.

[GAR 79] GARDNER N.J., HO P.T.J., "Lateral pressure of fresh concrete", *ACI Journal*, no. 76-35, p. 809-820, 1979.

[GAR 80] GARDNER N.J., "Pressure of concrete against formwork", *ACI Journal*, no. 77-31, p. 279-286, 1980.

[GHE 01] GHEZAL A., KHAYAT K.H., "Optimization of cost effective self-consolidating concrete", *Proceedings of the Second International Symposium on Self-Compacting Concrete*, p. 329-338, Japan, October 2001.

[HER 26] HERSCHEL W.H., BULKLEY R., "Konsistenzmessungen von Gummi-Benzollösungen", *Kolloid-Zeitschrift*, vol. 39, p. 291-300, 1926.

[KEM 69] KEMPSTER E., Pumpable concrete, current paper no. 29/69, Building Research Station, Garston, August 1969.

[KHA 98] KHAYAT K.H., "Viscosity-enhancing admixtures for cement-based materials – an overview", *Cement and Concrete Composites*, vol. 20, p. 171-188, 1998.

[KRI 59] KRIEGER M., DOUGHERTY T.J., "A mechanism for non-Newtonian flow in suspensions of rigid spheres", *Trans. Soc. Rheol*, vol. 3, p. 137-152, 1959.

[LAN 09] LANOS C., ESTELLÉ P., "Vers une réelle rhéométrie adaptée aux bétons frais", *EJECE*, vol. 13, no. 4, p. 257-471, 2009.

[LAR 96a] DE LARRARD F., BOSC F., CATHERINE C., DEFLORENNE F., "La nouvelle méthode des coulis de l'AFREM pour la formulation des bétons à hautes performances", *Bulletin des Laboratoires des Ponts et Chaussées*, no. 202, p. 61-69, 1996.

[LAR 96b] DE LARRARD F., SEDRAN T., HU C., SZITKAR J.C., JOLY M., DERKX F., "Evolution of the workability of superplasticized concretes: assessment with BTRHEOM rheometer", *Rilem International Conference on Production Methods and Workability of Concrete, RILEM Proceedings*, vol. 32, p. 377-388, Glasgow, Scotland, 3-5 June, 1996.

[LAR 98] DE LARRARD F., FERRARIS C.F., SEDRAN T., "Fresh concrete: a Herschel-Bulkley material", *Materials and Structures*, vol. 31, p. 494-498, 1998.

[LAR 00] DE LARRARD F., "Structure granulaires et formulation des bétons", *Ouvrage d'art OA 34*, Laboratoire Central des Ponts et Chaussés, Paris, April 2000.

[LAR 08] DE LARRARD F., SEDRAN T., "Betonlabpro: Une méthode scientifique de formulation des bétons", Laboratoire Central des Ponts et Chaussés, available at: www.lcpc.fr/fr/produits/betonlabpro, 2008..

[LEE 03] LEEMAN A., HOFFMAN C., "Pression of self-compacting concrete on the formwork", in O. WALLEVIK, I. NIELSSON (eds), *Proceedings PRO 33 of the 3rd International RILEM Symposium on Self-Compacting Concrete*, RILEM Publications, pp. 288-295, 2003.

[MAH 08] MAHAUT F., MOKEDDEM S., CHATEAU X., ROUSSEL N., OVARLEZ G., "Effect of coarse particule volume fraction on the yield stress and thixotropy of cementitious materials", *Cement and Concrete Research*, vol. 38, p. 1276-1285, 2008.

[MAT 00] MATHIEU D., PHAN-TAN-LUU R., "Planification d'expériences en formulation: criblage", *Techniques de l'Ingénieur, Génie des Procédés*, vol. 2, no. 240, Paris: Sciences et Techniques, p. 1-13, 2000.

[MAT 01] MATHIEU D., PHAN-TAN-LUU R., "Planification d'expériences en formulation: optimization", *Techniques de l'Ingénieur, Génie des Procédés*, vol. 2, no. 241, Paris: Sciences et Techniques, p. 2-3. 2001.

[OVA 06] OVARLEZ G., ROUSSEL N., "A physical model for the prediction of lateral stress exerted by self compacting concrete on formwork", *Materials and Structure*, vol. 39, no. 2, p 239-248, 2006.

[OZA 89] OZAWA K., MAEKAWA K., KUNISHIMA M., OKAMURA H., "Development of high performance concrete based on the durability design of concrete structures", *Proceedings of the second East-Asia and Pacific Conference on Structural Engineering and Construction (EASEC-2)*, p. 445-450, vol. 1, 1989.

[PER 09] PERROT A., AMZIANE S., OVARLEZ G., ROUSSEL N., "SCC formwork pressure: influence of steel rebars", *Cement and Concrete Research*, vol. 39, no. 6, p. 524-528, 2009.

[PRO 94] Projet National de recherche/développement Sablocrete, *Béton de sable: caractéristiques et pratiques d'utilisation*, Presses de l'ENPC, Paris, 1994.

[ROD 52] RODIN S., "Pressure of concrete on formwork", *Proceedings, Institution of Civil Engineers*, p. 709-746, vol. 1, part 1, no. 6, London, Great Britain, 1952.

[ROU 04] ROUSSEL N., LE ROY R., COUSSOT P., "Thixotropy modelling at local and macroscopic scales", *Journal of Non-Newtonian Fluid Mechanics*, vol. 117, p. 85-95, 2004.

[ROU 05a] ROUSSEL N., COUSSOT P., "Fifty cent rheometer for yield stress measurements: from slump to spreading flow", *Journal of Rheology*, vol. 49, no. 3, p. 705-718, 2005.

[ROU 05b] ROUSSEL N., LE ROY R., "The marsh cone: a test or a rheological apparatus?", *Cement and Concrete Research*, no. 35, p. 823-830, 2005.

[ROU 06a] ROUSSEL N., "A thixotropy model for fresh fluid concretes: Theory, validation and applications", *Cement and Concrete Research*, vol. 36, p. 1797-1806, 2006.

[ROU 06b] ROUSSEL N., COUSSOT P., "Ecoulement d'affaissement et d'étalement: modélisation, analyse et limites pratiques", *Revue Européenne de Génie Civil*, vol. 10, no. 1, p. 25-44, 2006.

[ROU 06c] ROUSSEL N., "A theoretical frame to study stability of fresh concrete", *RILEM Materials and Structures*, vol. 39, no. 1, p. 75-83, 2006.

[ROU 08] ROUSSEL N., OVARLEZ G., AMZIANE S., "Ecoulement et mise en œuvre des bétons", in N. ROUSSEL (ed.), *Collection Études et recherches des LPC, série Ouvrages d'art*, no. 59, 2008.

[SAA 01] SAAK A.W., JENNINGS H.M., SHAH S.P., "New methodology for designing self-compacting concrete", *ACI Material Journal*, vol. 98, no. 6, p. 429-439, 2001.

[SCH 00] SCHWARTZENTRUBER A., CATHERINE C., "La méthode du mortier de béton équivalent (MBE) – un nouvel outil d'aide à la formulation des bétons adjuvantés", *Materials and Structures*, vol. 33, p. 475-482, 2000.

[SVE 03] SVERMOVA L., SONEBI M., BARTOS P.J.M., "Influence of mix proportions on rheology of cement grouts containing limestone powder", *Cement Concrete and Composites*, vol. 25, no. 7, p. 737-749, 2003.

[TAT 73] TATTERSAL G.H., "The Rationale of a two-point workability test", *Magazine of Concrete Research*, vol. 25, no. 84, September 1973.

[TCH 08] TCHAMBA J.C., AMZIANE S., OVARLEZ G., ROUSSEL N., "Lateral stress exerted by fresh cement paste on formwork: laboratory experiments", *Cement and Concrete Research*, vol. 38, no. 4, p. 459-466, 2008.

[TOU 04] TOUTOU Z., LANOS C., MELINGE Y., ROUSSEL N., "Modèle de viscosité multiéchelle: de la pâte de ciment au microbéton", *Rhéologie*, vol. 5, p. 1-9, 2004.

[TUR 04] TURCRY P., Retrait et fissuration des bétons auto-plaçants: influence de la formulation, Thesis, Ecole Centrale de Nantes, 2004.

[WAL 03a] WALLEVIK O.H., "Rheology – a scientific approach to develop self-compacting concrete", *Proceedings of the 3rd International Symposium on Self-compacting Concrete*, p. 23-32, Reykjavik, Iceland, RILEM publications SARL, Bagneux, 2003.

[WAL 03b] WALRAVEN J., "Structural applications of self compacting concrete", *Proceedings of 3rd RILEM International Symposium on Self Compacting Concrete*, p. 15-22, Reykjavik, Iceland, RILEM Publications PRO 33, Bagneux, 2003.

[YAH 01] YAHIA A., KHAYAT K.H., "Analytical models for estimating yield stress of high performance pseudoplastic grout", *Cement and Concrete Research*, vol. 31, p. 731-738, 2001.

[YAH 03] YAHIA A., KHAYAT K.H., "Applicability of rheological models to high-performance cement grout containing supplementary cementitious materials and viscosity enhancing admixtures", *Materials and Structures*, vol. 36, no. 6, p. 402-412, 2003.

[YAM 07] YAMMINE J., Bétons fluides à hautes performances: relations entre formulation, rhéologie, physico-chimie et propriétés mécaniques, Thesis, ENS Cachan, 2007.

Chapter 2

Early Age Behavior

2.1. Introduction

Early age refers to the period which runs from when the concrete is cast until the onset of hardening. This period sees the material subjected to big changes, due to cement hydration: setting, heat release, deformations due to chemical reactions, and the evolution in mechanical properties. In addition to these phenomena, which originate endogenously, are those which are linked to the surroundings and to the dimensions of the structure being implemented: surface drying and deformations caused by desiccation, temperature increases and thermal deformations.

Studies on early age SCC have a primary focus on the characterization of the behavior in comparison with that of conventional concrete. SCCs behave *a priori* differently to conventional concretes, since their composition is different. The volumetric fraction of paste – a mixture of cement,

Chapter written by Philippe TURCRY and Ahmed LOUKILI.

mineral additions, water and additives – is higher for SCC: conventional concrete typically contains less than 30% paste whereas SCC generally contains more than 33% [DES 07]. The paste composition is also different for the two concrete types: the cement and water proportions are often close, but SCCs have higher mixture proportions of mineral additions and chemical admixtures.

In section 2.2 we show that early age behavior with regard to hydration and its consequences differs a little between SCCs and conventional concretes. In the following sections, which form the core of the chapter, we create a synthesis of the research on the consequences of early drying out of early age SCCs (plastic shrinkage and cracking).

2.2. Hydration and its consequences

SCCs contain higher proportions of mineral additions and additives than conventional concretes. These two composition parameters have an influence on cement hydration and its consequences.

2.2.1. *Hydration*

The influence of mineral additions on hydration depends on the nature of the additions. Limestone fillers are known to accelerate cement hydration in the short term, since they provide nucleation sites for hydrates [POP 05, BEN 06, BOU 08]. Note that in France, limestone fillers are the most frequently used additions for SCCs. The literature on fly ash, the combustion waste from power stations which is also used as a component of SCCs, is less clear cut regarding their influence on hydration. In general, a relatively low influence is observed [TUR 04, CRA 10].

Super-plasticizers tend to slow down cement hydration [TUR 04, HEI 07]. By bonding to the cement particles, the molecules of chemical admixtures slow the dissolving and diffusion of species in the interstitial solution [TOR 98].

2.2.2. *Setting*

Additives and mineral additions often have opposing effects on hydration: the water reducing additives slow it down while standard additives such as limestone fillers speed it up. As a consequence, SCCs display setting times of the same order, or a little slower, as those for conventional concretes fabricated with the same cement and W/C ratio [HEI 07].

2.2.3. *Chemical shrinkage and endogenous shrinkage*

Cement hydration is accompanied by a chemically-induced volume reduction known as Le Chatelier contraction [MOU 04]. This chemical shrinkage is of the same order of size for SCCs and conventional concretes made with the same cement, since it particularly depends on the chemical composition of the cement. The mineral additions and additives contained in the SCC paste affect the kinetics of the Le Chatelier contraction and have a small influence on its final size [BOU 08].

The effect of chemical shrinkage is macroscopic contraction which can be detected from the setting of a material protected from drying out, thus showing only endogenous shrinkage. Further, SCC behavior is a little different from that of conventional concrete. The W/C ratio for SCCs is, as a general rule, greater than 0.5, therefore too high for endogenous shrinkage to be important. Furthermore, the Le Chatelier contraction is detectably identical for the two types of concrete.

2.2.4. *Heat release, thermal contraction and the risk of cracking*

At the end of the setting, cement hydration results in a heat release which causes temperature increases, sometimes high, notably in large structures. These temperature changes are the cause of the gradient of thermal deformations between the skin and core of the concrete and can cause cracking. As for hydration, heat release is influenced by the proportions of mineral additions and super-plasticizer in the mixture [POP 05].

The risk of cracking has been studied by numerical simulations by [VOG 07]. The authors showed that the increase in temperature is a little faster for SCCs in the case of the most delicate structures. But the risk of cracking due to thermal deformations is paradoxically lower for SCCs. In fact, SCCs have, for the same compression resistance, a lower elastic modulus than conventional concretes (due to their higher proportion of paste, see Chapter 3). The stresses which develop after heat release are, as a consequence, lower.

2.3. Early age desiccation and its consequences: different approaches to the problem

When the surface of a concrete, whether fresh or during setting, is subjected to drying, the material contracts, since a reduction in capillary pressure develops in the porosity which is already saturated with water [WIT 76]. This phenomenon can be demonstrated in the laboratory using concrete samples of a few centimeters in size [KRO 95]. The macroscopic contraction, which is thus measured, is called the free plastic shrinkage.

At the level of a structure, concrete cannot deform freely: all material contraction is somewhat hindered. As a consequence, very early cracking in the concrete can result, even more rapidly since at this age the concrete has very low cohesion. In addition, drying is not uniform so the shrinkage gradient can also bring about a risk of cracking (Figure 2.1).

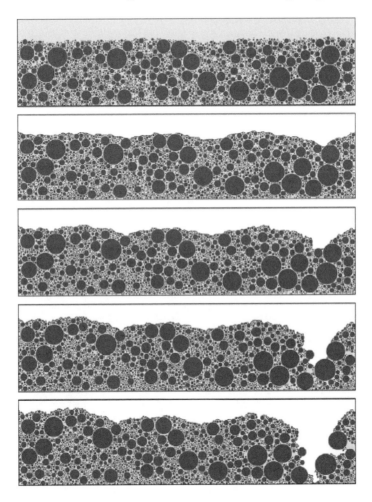

Figure 2.1. *Cracking of a granular region initially saturated with water, simulated in 2D. The granular distribution of particles modeled – 2 μm to 63 μm – is of the order of that of a cement. Cracking results from the effect of an increasing drop in capillary pressure (from top to bottom: 0 kPa, 24 kPa, 48 kPa, 72 kPa, 88 kPa) [SLO 09]*

Plastic shrinkage cracking can just be superficial as in the case of clay subjected to drying (Figure 2.2). However, this type of cracking degrades the skin of the concrete and therefore the covering which ensures the durability of reinforced structures.

Different approaches exist for studying the risk of cracking in fresh concrete. The most widely used approach consists of separating the problem by considering one part to be the response, i.e. the free plastic shrinkage, and the other part to be the deformation capability of the fresh concrete. The deformation capability is the maximum deformation of fresh concrete before cracking (Figure 2.3).

Figure 2.2. *Cracking caused by early drying out of a concrete slab [SLO 09]*

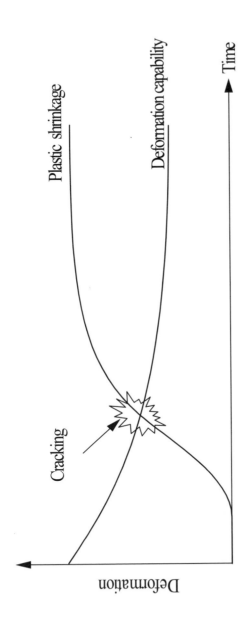

Figure 2.3. *Schematic diagram of the cracking mechanism of fresh concrete, in terms of competition between plastic shrinkage and deformation capability [BAY 02]. Plastic shrinkage develops in response to drying out, the slows down progressively since the material becomes rigid, notably during the setting. The deformation capability of the fresh concrete reduces, since the interstitial water is used up in hydration and drying, or even since the effect of a chemical additive such as superplasticizer diminishes over time*

Note that this approach to the risk of cracking is often limited to the measure of the free plastic shrinkage, since measuring the deformation capability is more difficult. The cracking risk is thought to increase with the free plastic shrinkage of the concrete. This method of viewing cracking is a little simplistic, since only the response is considered.

Another approach consists of measuring not the free shrinkage, but the drop in capillary pressure induced by drying. The risk of early cracking is considered to be higher in a concrete in which a drop in capillary pressure develops rapidly. In addition, it is without doubt a partial approach, since it only involves the response.

Finally, studying the risk of cracking can be done in a more direct fashion by conducting cracking tests on fresh concrete. With the aid of a device to prevent plastic shrinkage, usually with intense air circulation, it involves measuring the age at which cracks appear and quantifying the cracking by measuring the surface of the cracks produced at the end of the test. This type of test enables comparable studies of different compositions to be carried out. However, their representativeness can be questioned since the dimensions of the test samples and response are not very realistic.

2.4. Plastic shrinkage and drop in capillary pressure

2.4.1. Analysis of studied phenomena

2.4.1.1. Measuring apparatus

Different methods exist for measuring plastic shrinkage in the laboratory [KRO 95, HAM 98, BJO 99, FOU 00, HOL 00]. Here, we study the phenomenon on the basis of results obtained with the apparatus shown schematically in Figure 2.4 [TUR 04, TUR 06]. During the test, additional measurements to the shrinkage (horizontal deformation) are taken:

— monitoring the mass (required for determining the evaporation rate);

— temperature measurement (useful information for determining the end of the setting);

— settlement measurement (vertical deformation);

— measurement of the drop in capillary pressure using a pressure probe in the core of the concrete.

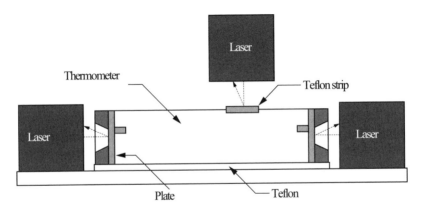

Figure 2.4. *Measuring apparatus for free plastic shrinkage and settlement using laser probes on a cuboid sample with dimensions 7x7x28 cm [TUR 06]*

2.4.1.2. *In endogenous conditions*

When the concrete surface is protected from drying out, the shrinkage measured during the first 24 hours is very low. The endogenous shrinkage, caused by hydration (auto-desiccation), is negligible, since the concrete mixtures considered here have a high W/C ratio (higher than 0.5).

2.4.1.3. *Controlled drying*

Figure 2.5 shows the shrinkage, settlement and mass loss curves for SCC subjected to controlled drying at 20°C and 50% relative humidity. Since the evaporation rate is constant (around 0.1 kg/m²/h), the changes in the kinetics seen in the deformation curves are therefore due to structural changes in the concrete.

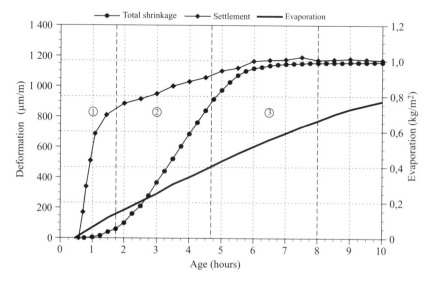

Figure 2.5. *Time-evolutions in shrinkage, settlement and evaporation of SCC subjected to controlled drying out at 20°C and 50% RH [TUR 04]*

During the first phase, the concrete experiences a high settlement, which is *a priori* the result of three phenomena: chemical shrinkage (or Le Chatelier contraction), volume reduction due to the loss of water by evaporation and consolidation under the effect of gravity. This last component of the settlement corresponds to a reduction in porosity which creates a rising of water to the surface or bleeding. During this first phase, the reduction in volume is transmitted gradually horizontally, partly because the drying caused a drop in capillary pressure (Figure 2.6), and partly because the internal frictions in the aggregate skeleton begin to be sufficient [TUR 04].

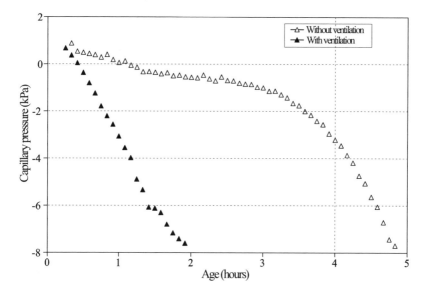

Figure 2.6. *Changes with time of the drop in capillary pressure in a SCC sample measured at 3.5 below the surface, during controlled drying "without wind" and forced drying "with wind" [TUR 04]*

During the second phase, shrinkage develops in a quasi-linear fashion with time. The speed of settlement becomes equal to the shrinkage speed: the volume reduction is therefore isotropic. The third phase corresponds to an inflection in the plastic shrinkage curve.

In the example given, the beginning and the end of the setting, measured by a Vicat test with the equivalent mortar method [SCH 00], correspond respectively to around 5 and 8 hours. The shrinkage curve is therefore inflected because the solid skeleton in formation acts bit-by-bit against the compression forces created by the drop in capillary pressure.

2.4.1.4. *Forced drying*

When air circulation (5 m/s) is applied to the sample surface, the evaporation rate increases to around 0.9 kg/m²/h. The deformation kinetics remain unchanged (Figure 2.7). On the other hand, the three phases observed in the previous case are brought forward in time. The final extent of the shrinkage is otherwise multiplied by about 4.

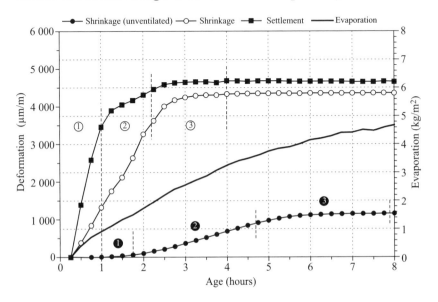

Figure 2.7. *Evolution of shrinkage, settlement and evaporation of SCC subjected to forced drying by air circulation at 5 m/s, at 20°C and 50% RH*

During the first phase, settlement results principally from evaporation. The drop in capillary pressure develops very rapidly (Figure 2.6). As a consequence, plastic shrinkage is detected from the beginning of the measurements. After this, the shrinkage evolves in a way which is quasi-linear with evaporation. During the third phase, we notice that, as in the case of controlled drying, the shrinkage curve is inflected. In the current case, the curve is inflected a long time before the beginning of the setting (estimated at

5 hours). The slow-down can therefore not be attributed to the development of rigidity in the solid skeleton due to hydration. This phenomenon is also observed in the case of soils subjected to drying: shrinkage slows down as it progresses, since the aggregate skeleton becomes compacted in such a way that it becomes rigid enough to resist all increases in confinement exerted by the drop in capillary pressure. We therefore speak of the shrinkage limit for soils.

2.4.1.5. *Synthesis*

After casting, the concrete deforms vertically under the effects of drying, Le Chatelier contraction and gravity. When the evaporation rate is very high, the shrinkage component linked to drying becomes dominant.

The kinetics of plastic shrinkage depends on the evaporation rate. Shrinkage occurs before the start of the setting when drying is forced. In the case of more controlled drying, the concrete contracts more slowly, before and during setting.

A good correlation is observed between plastic shrinkage and drop in capillary pressure until the solid skeleton becomes rigid enough to stay in the confinements caused by the drop in capillary pressure.

2.5. Comparison of plastic shrinkage for SCCs and conventional concretes

2.5.1. *Controlled drying*

Figure 2.8 shows plastic shrinkage for SCCs and conventional concretes made with a wide variety of constituents [TUR 04]. Each SCC is linked to a conventional concrete made from the same constituents and with the same compression resistance after 28 days. In controlled drying conditions (evaporation rate of around

0.1 kg/m²/h), SCCs systematically show plastic shrinkage of a higher magnitude than the conventional concretes. Furthermore, plastic shrinkage develops more rapidly in SCCs. In the literature, other results confirm this very marked tendency [GRA 99].

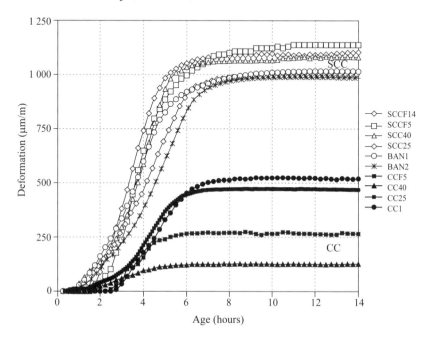

Figure 2.8. *Plastic shrinkage of SCC and conventional concretes (BO) subjected to controlled drying at 20°C and 50% RH [TUR 04]*

Three elements enable the differences in behavior between SCC and conventional concretes to be explained:

1. When the first phase of the phenomenon is studied (see section 3.2.1), conventional concretes subside more quickly and further than SCCs, as shown in Figure 2.9. The speed of settlement for SCCs is approximately equal to the speed of deformation calculated from the evaporation rate (Figure 2.10). From this, we deduce that the settlement measured for SCCs corresponds to the only volumetric reduction due to the

loss of water through drying out. Against this, the settlement speed for conventional concretes is much higher than that deduced from evaporation rates. This results from the measured settlement also being caused by the compaction of conventional concretes under their own weight. This consolidation necessarily favors the rising of water to the surface, i.e. bleeding.

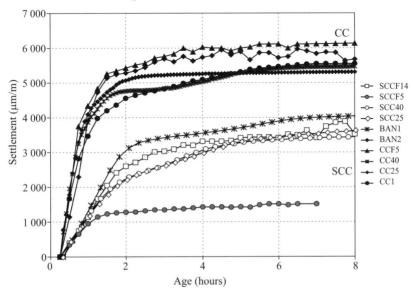

Figure 2.9. *Evolution of settlement in SCC and conventional concretes (CC) subjected to controlled drying at 20°C and 50% RH [TUR 04]*

Settlement measures suggest therefore that SCCs bleed much less than conventional concretes. Two composition parameters can explain this. First of all, SCCs have a higher volume of fine elements (cement and mineral additions) than conventional concretes (Figure 2.10). Several studies have also shown that bleeding is reduced when the proportion of super-plasticizer in the mixture is increased [JOS 02, TUR 04, HER 08].

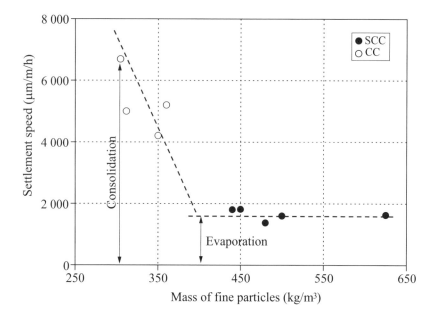

Figure 2.10. *Initial settlement speed as a function of the mass of fine particles in SCC and conventional concretes (CC) subjected to controlled drying at 20°C and 50% RH [TUR 04]*

2. The drop in capillary pressure develops more rapidly in SCCs, as shown by various results in the literature [HAM 03]. This is a consequence of the difference in bleeding: meniscuses appear more rapidly at the surface of SCCs. In addition, the pores at the surface of SCCs are *a priori* finer than those of conventional concretes, because of their low water/fines ratio. The drop in capillary pressure is inversely proportional to the pore radius, according to the Laplace relation.

3. When drying is controlled, the inflection in the plastic shrinkage is, as shown in section 3.2.1, linked to the setting of the materials. SCCs typically have settings that are a little longer than conventional concretes because they contain more super-plasticizer which tends to slow down

hydration. The period in which plastic shrinkage can develop is hence necessarily longer in the case of SCCs.

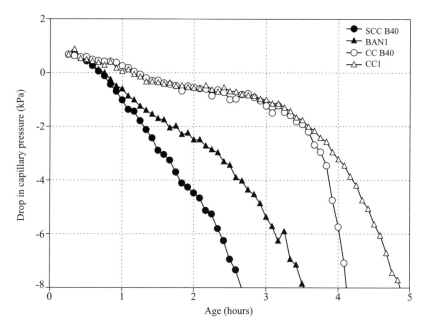

Figure 2.11. *Evolution of the drop in capillary pressure in SCCs and conventional concretes (CC) subject to controlled drying at 20°C and 50% RH*

2.5.2. *Forced drying*

When drying is forced by convection, a marked difference in no longer observed between SCCs and conventional concretes (Figure 2.12). On the contrary, SCC and a conventional concrete made with the same constituents have plastic shrinkage curves which are quite close. The final extent of the plastic shrinkage is however 10 to 20% higher for SCCs (in controlled drying, the shrinkage of SCCs is at least twice as high).

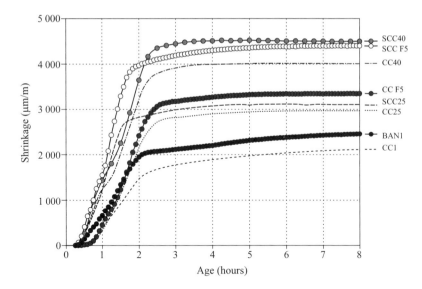

Figure 2.12. *Evolution of plastic shrinkage of SCCs and conventional concretes (CC) subjected to forced drying with air circulating at 5m/s, at 20°C and 50% RH [TUR 04]*

The three elements in the discussion above also serve to explain the observed behavior:

1. Bleeding is, with regard to plastic shrinkage, a natural cure for conventional concretes. When the evaporation rate is high, the bleeding water evaporates quickly from the surface. Conventional concretes are therefore not protected from drying out. The settlement speeds of the two types of concrete are, in any case, much closer than in the case of controlled drying (Figure 2.13).

2. Figure 2.13 shows an example of the development of the drop in capillary pressure with the same kinetics for SCC and a conventional concrete made with the same constituents. The response at the beginning of shrinkage is therefore almost identical for the two types of concrete.

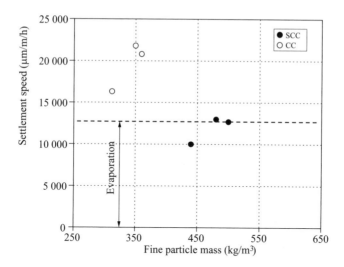

Figure 2.13. *Initial settlement speed as a function of the mass of fine particles in the SCC or conventional concrete (CC) subjected to a forced drying by a circulation of air at 5 m/s, 20°C and 50% RH [TUR 04]*

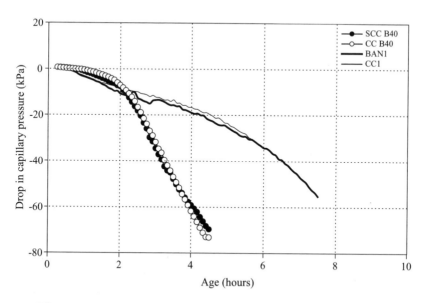

Figure 2.14. *Drop in capillary pressure of two pairs of SCCs and conventional concretes (CC) subjected to forced drying by a circulation of air at 5m/s, 20°C and 50% RH [TUR 04]*

3. For forced drying, plastic shrinkage largely occurs before setting, whatever the concrete type. The setting time therefore influences the extent of the shrinkage only a little, or not at all.

2.6. Influence of composition on free plastic shrinkage

2.6.1. *Influence of the paste composition*

Below are presented the principal results of a parametric study on mortars which have a fixed volumetric proportion of sand of 45% [TUR 04].

During controlled drying, the mixture proportion of the super-plasticizer is one of the parameters with the most affect on plastic shrinkage (Figure 2.15). The extent of plastic shrinkage increases with the proportion of super-plasticizer in the mixture, since sweating, which is correlated to the settlement speed, is reduced and the beginning of setting is delayed. When drying is forced by convection, the effect of the super-plasticizer on the shrinkage is more limited, and zero on the settlement speed (Figure 2.15).

A reduction in water (water/fines ratio) typically increases the extent of the shrinkage, but its effect is quite moderate. Of course, an earlier development of a drop in capillary pressure is favored (finer capillaries and more limited bleeding), but a reduction in water also enables more rapid setting to take place, which limits the development of shrinkage [WITT 76, KRO 97, HAM 98, HOL 00].

The proportions of mineral additions can also affect shrinkage, depending on the nature and grainsize of the addition used [TUR 04]. Plastic shrinkage also increases with finer addition particles (and finer cement particles) since the size of the capillary porosity is directly proportional to the fineness of the constituents.

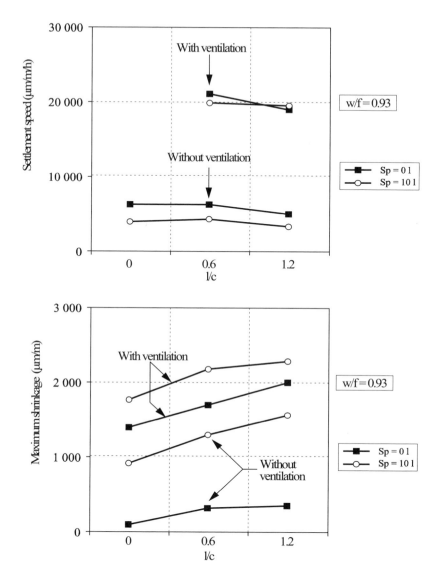

Figure 2.15. *Maximum extent of plastic shrinkage and settlement speed of mortars with different proportions of super-plasticizer (Sp) and ratios of limestone filler / cement (l / c) and water / fines (w / f) with or without forced convection*

2.6.2. *Influence of the paste proportion*

When drying is controlled, the paste proportion has little influence, as shown in Figure 2.16. It seems as though bleeding, the development of the drop in capillary pressure and setting are influenced very little by this formulation parameter.

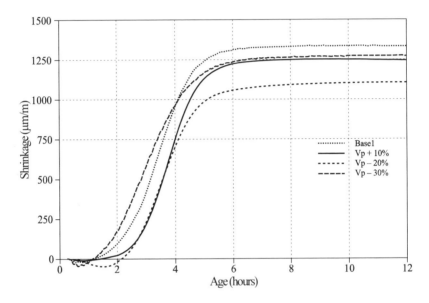

Figure 2.16. *Plastic shrinkage of concretes with 4 different volumetric proportions of paste (from -30 to +10% relative to the base composition) – controlled drying*

On the other hand, the influence becomes significant when air circulation is applied to the surface of samples: plastic shrinkage increases with the volume of paste (Figure 2.17). Up to the inflection point of the curves, the four compositions studied have the same shrinkage kinetics. The difference in final extent corresponds therefore to a difference in the shrinkage limit, like that observed for soils (see section 2.4.1.4). When the volume of paste is reduced, the rigidity of the aggregate skeleton increases more quickly with compaction of the material under the effect of the drop in capillary pressure. This phenomenon can be attributed to the effect caused by the aggregates becoming rigid and piling up.

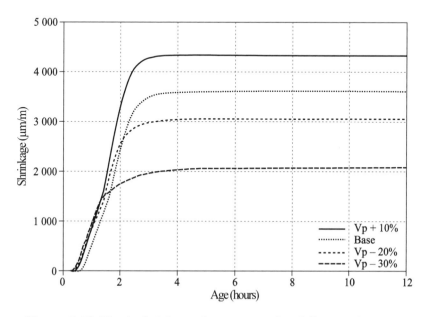

Figure 2.17. *Plastic shrinkage of concretes with 4 different volumetric proportions of paste (from -30 to +10% relative to the base composition) – forced drying*

2.7. Cracking due to early drying

Very few studies have been carried out on cracking in fresh SCCs. Here we present some of the results published by [TUR 06].

2.7.1. *Experimental apparatus*

Designed by Soroushian and Ravanbakhsh [SOR 98], the apparatus used is a rectangular mold with three triangular-profile raised parts at the bottom which initiate cracking

(Figure 2.18). The test is conducted in an air-conditioned room at 20°C with 50% relative humidity. A fan is set up next to the sample to force drying. Note that in the absence of forced ventilation, plastic shrinkage is not sufficient, regardless of the concrete type, to cause cracking. It is also very difficult to make small samples of a concrete, such as those presently used in laboratories, crack without forced drying.

Figure 2.18. *Mold for cracking test for fresh concrete*

During the test, the time at which the central crack appears is measured. At the end of the test, i.e. 8 hours after it began, the maximum width of the crack is measured using a graduated lens.

2.7.2. *Comparison of SCCs and conventional concretes*

When a SCC and a conventional concrete are made with the same constituents (for the same strength class), the SCC tends to crack less than the conventional concrete when fresh (Figure 2.19): cracking appears more slowly for the SCC and the central crack observed at the end of the test is narrower. The plastic shrinkage measured from forced drying is close for the two types of concrete (just higher for SCCs). Other elements must therefore be used to justify the differences in behavior.

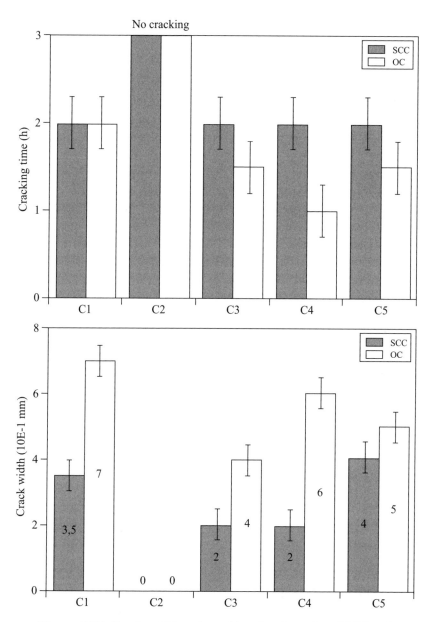

Figure 2.19. *Crack width and cracking time for pairs of SCC and conventional concretes (CC) with the same strength and constituents*

The first of such elements is the difference in deformation capability when fresh (see Figure 2.3). It seems that conventional concretes, which are necessarily much less fluid than SCCs, are also less deformable. Figure 2.20 shows the results of a cracking study of fresh mortars over a wide range of compositions and consistencies. It emerges from this program of cracking tests on fresh materials, with drying forced by a fan, that mortars which crack the least at a very early age are universally the most fluid mortars.

The greatest tendency in conventional concrete cracking is perhaps also the consequence of the apparatus used. In effect, the raised triangular sections at the bottom of the mold prevent plastic shrinkage, but also partially prevent settlement. Settlement is a little higher for conventional concretes (Figure 2.13).

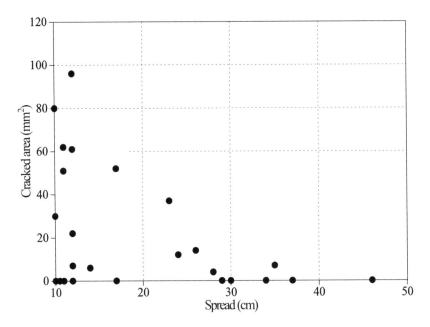

Figure 2.20. *Correlation between the total surface area of cracks measured at the end of cracking when fresh (with forced drying) and the spread of mortars with varied compositions*

2.8. Summary

Based on the data presented above, we are able to draw the following conclusions about shrinkage caused by early age drying and cracking risk.

When the evaporation rate is moderate (little air convection) plastic shrinkage of a concrete develops before and during setting. In these drying conditions, the extent of the shrinkage measured in the laboratory is much higher for SCCs than for conventional concretes. As shown by the settlement measurements, the difference in behavior between the two types of concrete is principally due to the difference in bleeding. While bleeding offers a natural protection against drying out for conventional concretes, bleeding hardly exists for SCCs because of their high proportions of fine elements and super-plasticizers. The drop in capillary pressure, which leads to the evaporation of interstitial water, develops more rapidly for SCCs.

When the rate of evaporation is high (drying with convection), plastic shrinkage of a concrete occurs mainly before setting. In these conditions, SCCs and conventional concretes have similar extents and kinetics of plastic shrinkage, since the principal causes of the observed differences in behavior with moderate drying disappear. In these drying conditions, fresh concrete cracking tests show that SCCs tend to crack less than conventional concretes. It seems that the tendency of SCCs to crack less is, in this case, due to their greater deformation capacity when fresh.

In reality, a cure for SCCs, which prevents the evaporation of interstitial water after casting, is to recommend compensation for not bleeding and thus reduce the risk of early cracking. Comparison of SCCs and conventional concretes also shows the need, of all types of concretes, for protection against drying out at an early age, in particular when the evaporation rate is high.

The tendencies presented here on the fresh state behavior really depend on the chemical admixtures used in large quantities in SCC mixtures. In the future, these tendencies will, perhaps, be modified, since the molecules used to fluidize SCCs continue to evolve.

That being the case, further research could be carried out in order to better understand the influence of drying at an early age on the cracking risk by restrained shrinkage in both the short and long term. It would be interesting to develop a cracking test for concretes when fresh or during setting for moderate evaporation rates, since, in these drying conditions, studies in the literature are only concerned with "responses" (free shrinkage or drop in capillary pressure).

The cracking mechanism in fresh concrete is complex because the material changes significantly. When fresh, concrete is a granular material, almost a suspension or a saturated powered soil, in which cohesion develops as a result of hydration. From a visco-elastic regime, it moves during setting into an elasto-plastic regime. Modeling will be, without doubt, welcomed for a better understanding of the phenomenon.

Until now models have focused either on fresh concrete modeled as a granular regime (with discrete models for example [SLO 09]), or on concrete after setting has begun – a regime which evolves into being elasto-plastic [MES 06]. As far as we know, no model yet links before and after setting.

2.9. Bibliography

[BAY 02] BAYASI Z., MC INTYRE M., "Application of fibrillated polypropylene fibres for restraint of plastic shrinkage cracking in silica fume concrete", *ACI Materials Journal*, p. 337-344, 2002.

[BJO 99] BJONTEGAARD O., Thermal dilation and autogenous deformation as driving forces to self-induced stresses in high performance concrete, PhD Thesis, Norwegian University of Science and Technology, Trondheim, 1999.

[BOU 08] BOUASKER M., MOUNANGA P., TURCRY P., LOUKILI A., KHELIDJ A., "Chemical shrinkage of cement pastes and mortars at very early age: Effect of limestone filler and granular inclusions", *Cement Concrete Comp*, vol. 30, no. 1, p. 13-22, 2008.

[CRA 10] CRAEYE B., DE SCHUTTER G., DESMET B., VANTOMME J., HEIRMAN G., VANDEWALLE L., CIZER O., AGGOUN S., KADRI E.H., "Effect of mineral filler type on autogenous shrinkage of self-compacting concrete", *Cement and Concrete Research*, 2010.

[DES 07] DE SCHUTTER G., BOEL V., "Self-Compacting Concrete SCC", *Proceedings of the Fifth International RILEM Symposium on SCC*, Ghent, Belgium, 2007.

[FOU 00] FOURDIN E., GUIGOU C., CHAPPUIS J., "Early age shrinkage of mortars: conception of a new device and detailed analysis of a typical experimental curve", *Proceedings of the International RILEM Workshop, Shrinkage 2000*, RILEM Publications, Paris, 2000.

[GRA 99] GRAM H.E., PIIPARINEN P., "Properties of SCC – especially early age and long term shrinkage and salt frost resistance", *Proceedings of the First International RILEM Symposium of Self-Compacting Concrete*, Stockholm, Sweden, 1999.

[HAM 98] HAMMER T.A., "Cracking in high performance concrete before setting", *Proceedings of the International Symposium on High Performance and Reactive Powder Concrete*, Sherbrook, Canada, 1998.

[HAM 01] HAMMER T.A., "On the strain capacity and cracking mechanisms of HSC at very early age", *Proceedings of the 6th International Conference Concreep*, Cambridge, USA, 2001.

[HAM 03] HAMMER T.A., "Cracking susceptibility due to volume changes of self-compacting concrete", *Proceedings of the Third International RILEM Symposium of Self-Compacting Concrete*, Reykjavik, Iceland, 2003.

[HEI 07] HEIRMAN G., VANDEWALLE L., VAN GEMERT D., "Influence of mineral additions and chemical admixture on setting and volumetric autogenous shrinkage of SCC equivalent mortar", *Proceedings of the Fifth RILEM Symposium on SCC*, Gand, Belgium, 2007.

[HOL 00] HOLT E., LEIVO M., "Methods of reducing early-age shrinkage", *Proceedings of the International RILEM Workshop Shrinkage*, Paris, 2000.

[KRO 95] KRONLOF A., LEIVO M., SIPARI P., "Experimental study on the basic phenomena of shrinkage and cracking of fresh mortar", *Cement and Concrete Research*, vol. 25, no. 8, p. 1747-1754, 1995.

[MES 06] MESSAN A., Contribution à l'étude du comportement au très jeune âge des structures minces en mortier, University of Montpellier II, 2006.

[SLO 09] SLOWIK V., HUBNER T., SCHMIDT M., VILLMANN B., "Simulation of capillary shrinkage cracking in cement-like materials", *Cement and Concrete Composites*, vol. 31, p. 461-469, 2009.

[SOR 98] SOROUSHIAN P., RAVANBAKHSH S., "Control of plastic shrinkage cracking with specialty cellulose fibres", *ACI Materials Journal*, vol. 95, no. 4, p. 429-435, 1998.

[TOR 98] TORRENTS J.M., RONCERO J., GETTU R., "Utilization of impedance spectroscopy for studying retarding effect of a superplasticizer on the setting of cement", *Cement and Concrete Research*, vol. 28, p. 1325-1333, 1998.

[TUR 04] TURCRY P., Retrait et fissuration des bétons auto-plaçants, influence de la formulation, PhD Thesis, Ecole Centrale de Nantes – University of Nantes, 2004.

[TUR 06] TURCRY P., LOUKILI A., "Evaluation of plastic shrinkage cracking of self-compacting concrete", *ACI Material Journal*, vol. 103, p. 272-279, 2006.

[WIT 76] WITTMAN F.H., "On the action of capillary pressure in fresh concrete", *Cement and Concrete Research*, vol. 6, no. 1, p. 49-56, 197.

Chapter 3

Mechanical Properties and Delayed Deformations

3.1. Introduction

This chapter aims to produce a synthesis of the research carried out on the behavior of SCCs in the hardened state, i.e. after setting. Instantaneous mechanical properties and delayed properties of SCCs are of interest, in comparison with those of conventional concretes in the same class of compressive strength by identifying the influence of mix-design parameters. A second aspect is concerned with the standards dealing with the design of structural concrete in order to validate or adapt the codes to these materials, a condition *sine qua non* in the widespread adoption of their use.

Comprising a wide variety of material formulations, the data presented in this chapter come from procedures and experimental apparatus which are sometimes very different

Chapter written by Thierry VIDAL, Philippe TURCRY, Stéphanie STAQUET and Ahmed LOUKILI.

from one another. In any case, researchers have described their comparative studies of conventional concretes and SCCs by fixing various parameters (whether compressive strength, W/C ratio, or otherwise). These details can lead to sometimes contradictory results. We endeavor to propose a universal analysis by clarifying the factors which could be at the root of sometimes opposing conclusions. Comparisons with design rules and real calculations are both presented in order to verify their precision in the case of SCCs.

3.2. Instantaneous mechanical properties

Describing the instantaneous mechanical behavior of SCC involves defining its principal characteristics, knowing its compressive strength, its elastic modulus in compression and its tensile strength (either by a direct tension test or by splitting). These properties are crucial since they form the basis for the specification rules for reinforced or prestressed concrete.

3.2.1. *Time-evolution of compressive strength*

3.2.1.1. *Phenomenon studied and analysis method*

Studies of the kinetics of compressive strengths are generally carried out by normalizing the compressive strength relative to its value at 28 days. This analysis method is justified by the possible differences between the compressive strength of concretes initially formulated to aim for a particular strength class. In effect, if a comparison is made between SCC and an conventional concrete with the same W/C ratio, the reactivity of the mineral addition, which varies according to its type (limestone filler, fly ash, etc.), generally leads to a higher compressive strength in SCC, with a value that is close enough to that of conventional concrete that they can be considered to belong to the same strength class.

3.2.1.2. *Tendencies and influencing factors*

Research on the time evolution of compressive strength of SCCs has shown that it is universally similar to that of conventional concretes in the same strength class. No significant difference has been detected in the period before maturation and the end of the cement's hydration which is conventionally considered to occur at 28 days [VIE 03, TUR 06, ASS 03].

However, two specific details in SCC mix-design can give rise to small modifications in the kinetics, but with opposite effects:

– Additives: a propensity to develop delays in setting at an early age under the effect of adding additives has been observed, relative to their proportions in the mixture [PON 03] or their type [KHA 98a, BOS 03].

– Mineral additions: the incorporation of mineral additions also seems likely to modify hydration kinetics in the cement matrix. Notably, limestone fillers currently in use have the effect of accelerating hydration at early ages [PER 99, BON 00]. This results in a reduction in the start and end times of setting. Above all, mineral addition particles act as a nucleation site for C-S-H. This effect is all the more pronounced when the proportion of mineral additions is significant or their particles are finer. In any case, a minor proportion of filler can react with aluminates (C$_3$A) to form carbo-aluminates. These reactions are obviously affected by the C$_3$A content of the cement (see Chapter 2).

By the end of 28 days, the concrete must show a non-negligible increase in its compressive strength, which can reach 20% by 90 days [MAZ 05] under the effect of pozzolanic reactions of the mineral additions such as in the case of fly ash [MNA 08].

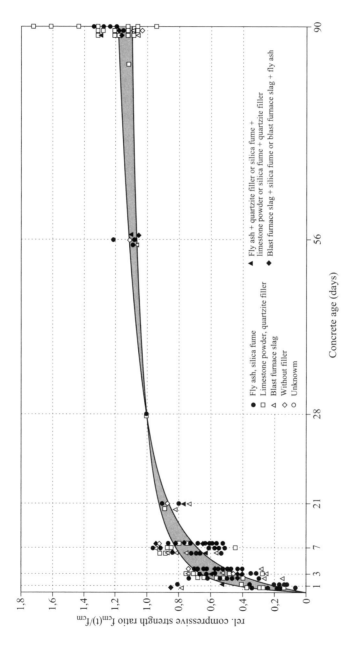

Figure 3.1. *Kinetics for compressive strength ($f_{cm}(t)$ / f_{cm} ratio) for SCCs made with different fillers and compared with predictions from CEB Model Code 90 [KLU 03]*

3.2.1.3. *Standard aspects*

Standard models, notably the CEB model [CEB 90] found in Eurocode 2, are still useful provided that the kinetics have little variation from those of conventional concretes (Figure 3.1) [KLU 03, MAZ 07]. In the presence of pozzolanic mineral additions, the increase in strength after 28 days is not taken into account, but this under-estimation stays safe.

3.2.2. *Tensile strength*

3.2.2.1. *Property studied and analysis method*

Tensile strength can be measured by a direct tension test, which is difficult to conduct, and by a splitting test (also known as a "Brazilian test"). In the rules for calculating structures, this is deduced from the unique value of the compressive strength.

3.2.2.2. *Tendencies and influencing factors*

Results concerning tensile strength are somewhat contradictory. Nevertheless, it is important to note that the differences in values between SCCs and conventional concretes are reasonable. Some studies [BOS 03, TUZ 06] did not distinguish any difference between the tensile strengths of the two types of material, if uncertainty and accuracy of the measurements is taken into account, or if it is argued to be the same as the compressive strength in order to eliminate variations in compressive strength between SCCs and conventional concrete [CAS 09].

[KON 03] and [SON 99] observed even better tensile strengths for SCCs, attributed to better homogeneity of SCCs. [DIN 07] linked this increase in the maximum tensile strain to the greater deformability of SCC paste and to the higher quantity of paste. [KLU 03] confirmed this observation and explained it as a resistance to tensile strain transfer which is improved by the quality of adhesion at the paste-aggregate interface level (ITZ – *interfacial transition zone*).

On the other hand, lower strengths have been observed in other research, which have reached a maximum value of 18% [PAR 07]. The hypotheses trying to identify the root of these results put the high quantity of filler used and the type of super-plasticizer first. In the presence of polycarboxylate super-plasticizer, [RON 02] detected the large crystals of portlandite and ettringite which can make the ITZ fragile and give rise to a reduction in tensile strength at the macroscopic level of the material.

3.2.2.3. *Statutory aspects*

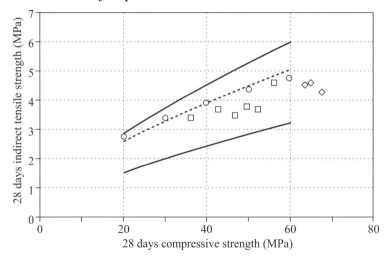

Figure 3.2. *Tensile strength measure by splitting as a function of compressive strength for different SCC formulations and compared with the spindle from CEB Model Code 90 [FEL 07]*

Tensile strength is narrowly linked to the compressive strength value and empirical relations can be found between the different standards. [FEL 07] observed that their experimental data, when completed by those published in the literature, are correctly bounded by the upper and lower limits according to the CEB-FIB model in Eurocode 2 (Figure 3.2).

The comparisons conducted by [KLU 03] are in agreement. One notable point is that the incorporation of silica fumes as an addition leads to significantly higher strengths in the higher region of this very model. Hence for a high performance SCC containing silica fumes and a target class of C80, the average tensile strength by cracking on 11/22 cylinders reaches 6.1 MPa after 28 days with a compressive strength of 116 MPa on 16/32 cylinders [STA 07].

3.2.3. *Elastic modulus*

3.2.3.1. *Property studied*

The elastic modulus in compression is a key parameter in design models especially for studies in the serviceability limit state (SLS) which is based on the behavior of materials in their elastic regime. Some aspects in the mix-proportions of SCCs raise questions regarding the value of this mechanical property for SCCs relative to conventional concretes.

3.2.3.2. *Tendencies and influencing factors*

If literature results are summarized, the almost unanimous conclusion is that the elastic modulus for SCCs is lower than that for conventional concretes in the equivalent strength class but only by a small proportion [PON 03, KON 03, CHO 03, KLU 03, TUR 06]. Some researchers have confirmed that the elastic moduli can be considered to be equivalent, if uncertainties in measurements are taken into account [PER 01, ASS 03]. The variation in the differences usually comes from different measuring procedures following the standards used and very different formulations. Comparisons of elastic moduli are sometimes carried out based on the reasoning that with the same W/C ratio in SCC and conventional concrete, detectable differences may arise in the compressive strengths which result from the effects of mineral additions, as discussed above. [DOM 07] collected the results of several studies (Figure 3.3) linking compressive strength (measured on a cube) to the elastic modulus for SCCs and conventional concretes. Analysis of this figure shows that the results are dispersed but also confirms that in general, SCCs are more deformable.

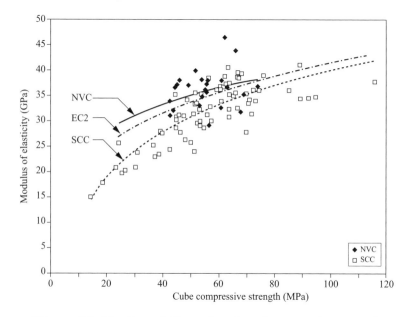

Figure 3.3. *Elastic moduli as a function of compressive strength (measured on cubes) for different SCC and conventional concrete formulations, compared with Eurocode 2 (EC2)* *[EUR 06] [DOM 07]*

Two influential parameters which are agreed upon by the research community are the coupled factors of the paste volume (mixture of water, cement, filler and additives) and the aggregate content. As for all composite materials, this characteristic depends on the value of each constituting phase. In the case of a concrete, aggregates are more rigid than cement paste. This results in a elastic modulus which is higher when the cement matrix volume is lower [KON 03, CHO 03, PER 05, VIE 03].

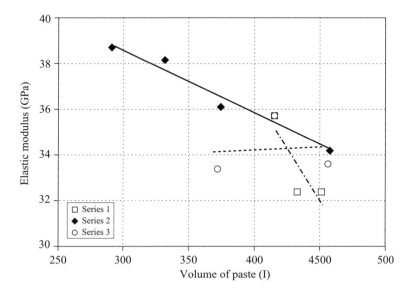

Figure 3.4. *Evolution of the elastic moduli for SCC formulations as a function of paste volume [ROZ 07]*

Rozière *et al.* [ROZ 07] studied several series of SCC mixtures with different paste volumes in order to determine the impact of this parameter (Figure 3.4). Series 1 shows a reduction in the elastic modulus associated with the paste volume with a constant W_{eff}/C ratio. However, they estimated that the effect of this parameter remains limited since the modulus value only decreases by 14% when the volume increases by 57%. Furthermore, the evolution in the same fashion of the compressive strength from 47.1 to 41.7 MPa must be considered. In the case of series 2, the proportion of water increases to the detriment of the aggregate content and the W_{eff}/C ratio, and produces the same effect. Finally, increasing the quantity of limestone filler in the mixture, combined with the lower proportion of aggregates, at the same W_{eff}/C ratio for series 3, does not show a confirmed tendency like the two previous series. The high filler content improves the strength of the ITZ and enables a slight increase in the compressive strength by a little activity, but this effect of increasing the rigidity is thwarted by the low proportion of aggregates.

On the basis of a study of mechanical properties of several SCCs with different quantities of fly ash (as a % of cement), [DIN 07] showed that the differences between the moduli of ordinary control concretes and SCCs are first and foremost significant for the lower strength classes, with strong substitution rates of cement by fly ash. These differences tend to disappear with high performing concretes for classes above 60 MPa [PAR 07].

Finally, the maximum grain size parameter for the aggregates which is reduced in SCCs in order to improve flow properties in the fresh state by influencing friction phenomena, has also been described to partly explain the high deformability of SCCs [JAC 99]. As in the case of compressive strength, the moduli of elasticity at long term can be higher than those for ordinary concretes [VIE 03, MAZ 05]. In fact, cement hydration is often favored by the lower loss of mass due to drying for SCCs. Besides, pozzolanic effects from the incorporated mineral addition can also increase SCC rigidity.

3.2.3.3. *Standard aspects*

Conventionally and according to standards, the elastic modulus is mainly correlated with compressive strength [RUL 00, BPE 99, EUR 06, ACI 02, CEB 90]. Comparison of experimental measurements has demonstrated a slight under-estimation on the part of the measurement standards, between 0 and 10% of the American standard ACI 318 [SCH 07].

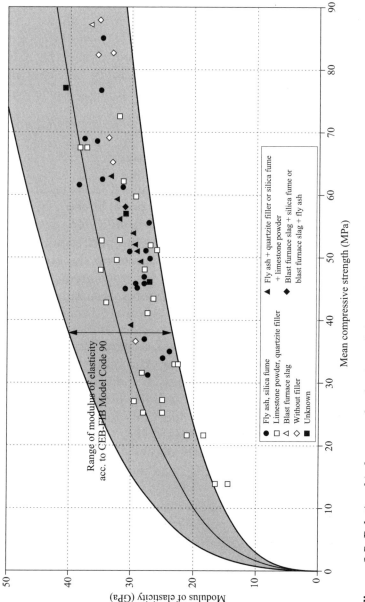

Figure 3.5. *Relationship between the elastic modulus and compressive strength for different SCC formulations (different filler types) and comparison with predictions from the CEB Model Code 90 [KLU 03]*

The differences remain low with predictions from Eurocode 2, and also with BAEL as observed by Pons *et al.* [PON 03]. Figure 3.5 [KLU 03], which shows the experimentally measured moduli for SCCs, formulated with various additions and values calculated using the CEB model, underlines the greater deformability of SCCs but nevertheless sufficiently moderate that Standard models are still appropriate.

3.3. Differences in mechanical behavior

Knowledge of delayed deformation of a cementitious material, i.e. free "shrinkage" deformations, "creep", is, as for the instantaneous behavior, indispensable for designing reinforced concrete structures. These properties allow the final deflections (maximum displacements) of the structural elements and prestress losses to be estimated, and the redistribution of strain in hyperstatic structures to be calculated.

As has been previously described, a significant volume of paste in SCCs is necessary to obtain the required properties in the fresh state. This mix-design parameter tends to increase the instantaneous deformability of the material, but this is sufficiently limited in proportions to allow the use of current predictive models. However, some long-term deformations which significantly exceed these values can be prejudicial to the delayed SCC behaviors (excessive flexing, risk of cracking, etc.).

To address this key investigation, numerous studies concerning shrinkage, restrained shrinkage and creep have been carried out on SCCs in order to determine their long-term behavior and the influential parameters.

3.3.1. *Free shrinkage*

3.3.1.1. *Property studied and characterization*

Shrinkage corresponds to a collection of phenomena which lead to contraction in the cement matrix in the absence of external mechanical strain. We are interested here only in shrinkage in hardened concretes, rather than in that which occurs before setting, notably plastic shrinkage which must not be neglected since it can be the cause of early cracking (Chapter 2). There are two types of shrinkage which are usually considered separately.

We speak of autogenous shrinkage if no exchange of water is allowed with the ambient surroundings (for example behavior inside a very large structure), which is itself associated with different shrinkage mechanisms:

– Le Chatelier contraction due to the lowest volume of hydrates formed compared to the initial volume of water and anhydrous cement;

– thermal shrinkage linked to variations in temperature which arise as a consequence of exothermic hydration reactions;

– auto-desiccation shrinkage which begins when the skeleton forms after the cement has set and which is due to a drop in capillary pressure which occurs in the pores that have been emptied of water during hydration (the extent of contraction is inversely proportional to pore size, following the Kelvin-Laplace law).

If water exchanges are possible, shrinkage known as desiccation shrinkage or drying out shrinkage occurs in addition to the autogenous shrinkage (for example behavior at the free surfaces of the structure) caused by the water gradient between the material and the surroundings. The sum of these two shrinkages is called the total shrinkage.

Various measuring devices exist, notably for autogenous shrinkage. In general, measurements are recorded one day after concrete fabrication. The variability in measurement procedures can sometimes be taken into account to explain disagreements in the conclusions from studies in the literature. Hence Staquet *et al.* [STA 06] showed that, from when the concrete is cast, a high performance SCC develops, in less than 48 hours, more than 50% of its autogenous shrinkage at 40 days.

In this study, the zero time that corresponded to when the shrinkage began started at the end of the concrete's setting, otherwise said to be less than 1 day. This means that a non-negligible proportion of shrinkage can be hidden by following measurement techniques.

3.3.1.2. *Tendencies and influencing factors*

Research carried out on shrinkage can be classified into two groups. The first group comprises most of the authors [OGA 95, ROL 99, TAN 93, PER 01, TUR 06, FER 07, ROZ 07] who conclude that SCC shrinkage, whether autogenous or due to desiccation, is of the same order of magnitude, or a little higher with little difference, as those of conventional concretes of the same strength class (Figure 3.6).

Other researchers [PON 03, VID 05] determined that they did not distinguish a particular behavior for SCCs, on the condition that the basic concrete components, excluding additions, of course, are identical. On the other hand, a second group of researchers observed more significant differences [KLU 03, XIE 05, CHA 04, PER 05, MAZ 05, HEI 08], with the values being up to 20% higher in the case of SCCs [CHO 03].

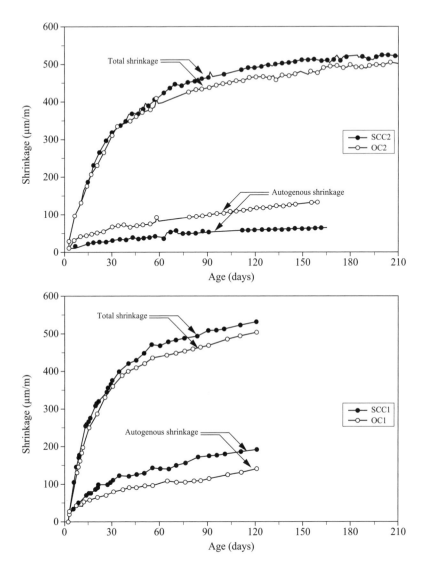

Figure 3.6. *Comparisons of changes in autogenous shrinkage and total shrinkage for two SCCs and two conventional concretes in the same strength class and made from the same constituents [TUR 06]*

Besides measurement methods and precision, test conditions seem to play a role in varying the results.

However, mix-design criteria can explain shrinkages which are sometimes the same or sometimes larger. Several parameters, specific to SCC mixtures and which have coupled influences, have been identified:

– Paste volume is the most significant factor, since shrinkage develops in the cement matrix. Hence, after [ROZ 07], total shrinkage increases linearly with paste volume at constant W/C and filler/C ratios.

– Mineral addition proportions, which make the matrix denser while reducing the porosity and the pore size, can cause more intense autogenous shrinkage. Hence, Staquet *et al.* [STA 06] showed that a high performance SCC with silica fumes develops an autogenous shrinkage deformation of more than 100 μm/m in less than 48 hours after setting.

– Type of mineral additions: with identical W/C and filler/C ratios, and the same quantity of water and aggregate, SCCs formulated with different filler types (calcium, dolomite, fly ash, quartz filler) develop different shrinkages [HEI 03]. The largest deformations are obtained when limestone filler is characterized with the greatest fineness, a factor which seems influential [KLU 03].

– Maximum aggregate size: the bigger the aggregate, the smaller the shrinkage. This phenomenon is obviously undesirable with regard to the requirements that SCCs must satisfy in the fresh state.

Other factors have an important impact, such as the W/C ratio which controls porosity and compressive strength. A low W/C ratio tends to intensify autogenous shrinkage since smaller pores are formed, but reduces desiccation shrinkage. These are not, however, specific to SCCs and their effect remains in evidence in a similar way for conventional concretes.

3.3.1.3. *Standard aspects*

Since standard models have been established based on correlations which come from a dataset composed of results obtained from conventional concretes, they tend to under-estimate shrinkage deformations for SCCs in proportions which are, however, reasonable.

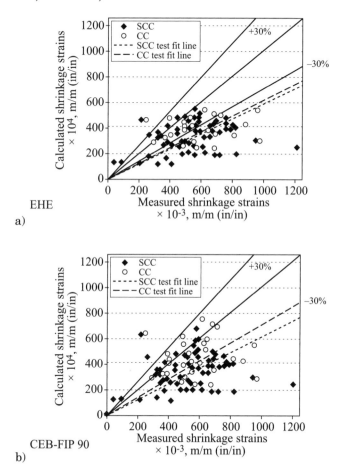

Figure 3.7. *Comparison of experimental shrinkage values from the literature with values calculated according to different standard models or proposed by researchers for conventional concretes (CC) and SCCs [FER 07]*

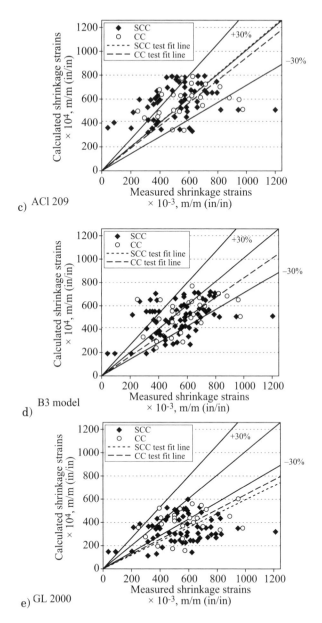

Figure 3.7. (continued) *Comparison of experimental shrinkage values from the literature with values calculated according to different standard models or proposed by researchers for conventional concretes (CC) and SCCs [FER 07]*

Comparison of results from a summary of several research works with various normative predictive models or proposed by researchers [FER 07] illustrates this global tendency (Figure 3.7). In this figure, the results are presented in terms of calculated shrinkage strains as a function of *measured shrinkage strains*. The bisector at 45° shows the equivalence between these deformation values and the two other lines which correspond to the generally admitted limits of uncertainty at ± 30%.

3.3.2. *Restrained shrinkage*

3.3.2.1. *Phenomenon studied and characterization*

Shrinkage, when it occurs in structures which are subjected to conditions which prevent them from deforming freely, can induce tensile stress which can cause cracking when the tensile strength of the material is reached. This phenomenon is therefore detrimental to structure durability.

Susceptibility to this type of cracking is usually evaluated using a ring test (Figure 3.8). The principle of the test relies on indirectly monitoring the deformation of a ring-shaped test sample until it cracks, the deformation being blocked at the inner surface of the ring by a steel ring [TUR 06, ROZ 07], such that the radial surfaces are subjected to drying out [HWA 10] complying with the [AST 05] standard.

Using these results, a cracking potential can be defined from the direct measurement of the time at which cracking occurs, and from parameters calculated from the cracking mechanism such as the cracking energy, characteristic parameters for cracking (CTOD, *critical crack tip*, Kic tenacity), the tensile creep coefficient, etc.

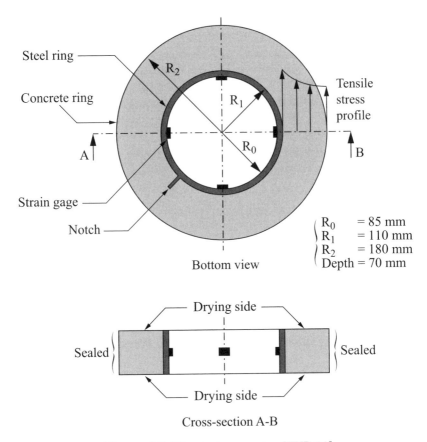

Steel ring

Concrete ring

R_2

R_1

R_0

A

B

Tensile
stress
profile

Strain gage

Notch

Bottom view

R_0 = 85 mm
R_1 = 110 mm
R_2 = 180 mm
Depth = 70 mm

Drying side

Sealed

Sealed

Drying side

Cross-section A-B

Figure 3.8. *Ring test apparatus [TUR 06]*

3.3.2.2. *Tendencies and influencing factors*

Investigations on restrained shrinkage are still rare since the codes for reinforced concrete structure design do not suggest any explicit relations which take this phenomenon into consideration. From a few studies carried out on the subject, it seems that the cracking potential for SCCs is, on average, slightly higher than for conventional concretes [TUR 06, HWA 10, LOS 09], and notably characterized by a very early cracking appearance age. The cracking risk is also higher when the SCC's stability is too low.

More generally, this behavior is linked to a slightly higher desiccation shrinkage for SCCs, owing principally to the higher paste volume which increases the strain limits and on average reduces the cracking age [ROZ 07]. However, other interacting factors must also be taken into account such as the elastic modulus and the tensile strength which are not affected in the same way by the paste volume.

The additive type also turns out to be a determining factor. The use of a highly water reducing polynaphthalene sulfonate type of super-plasticizer doubles the cracking age in comparison with a polycarboxylate super-plasticizer [HWA 10]. Finally, these researchers have shown how the cement type (with additives such as silica fume, fly ash, slag, etc.) affects the cracking strength for restrained shrinkage, at the same W/C ratio.

3.3.3. *Evolution and prediction of delayed deformations under loading, creep deformations*

3.3.3.1. *Property studied and characterized*

Creep deformations correspond to delayed deformations which occur under a mechanical stress which is sustained over time. Since concrete is a material used to support compressive forces, its creep behavior is naturally and principally studied under uniaxial compression conditions. The origin of these viscous deformations is strongly related to movements of water in the paste (by diffusion) in the short term, and slipping layers of C-S-H in the longer term.

As in the case of shrinkage, the total creep, evaluated in the desiccation mode, corresponds to the sum of what is called "basic creep" in an autogenous mode (without water exchange) and of what is called "desiccation or drying creep" which is due to the mobilization of a complementary desiccation shrinkage which results from mechanical responses, this differentiation is purely arbitrary.

Creep deformation is calculated by taking away, also arbitrarily, the instantaneous and shrinkage deformations from the total deformation under load. The results can be expressed as specific creep or compliance, by relating the creep deformation to the creep stress applied. This enables comparison of creep for concretes from varied strength classes with values normalized by units of strain.

Creep strain corresponds to a percentage of the concrete's compressive strength at the time at which it is loaded. Finally, a creep coefficient can also be defined, a deformation ratio of creep/instantaneous deformation. This is often used since it is a parameter which is taken into account in the design codes.

3.3.3.2. *Tendencies and influencing factors*

Synthesis of the comparative studies on creep between SCC and conventional concrete are similar to those on shrinkage. Creep deformations are slightly higher with differences which can be considered negligible (Figure 3.9) [VID 05, PON 03, PER 05, CHO 03, VIE 03, TUR 06] or more important [ASS 03, MAZ 05, HEI 08].

The mode in which the creep results are used and expressed, multiple experimental uncertainties linked to measuring precision (measuring apparatus with probes, gauges, etc.), scaling effects (sample geometry), expiry of the end of recording values, as well as the ambient conditions on occasion, can partly explain these small differences.

The principal parameter of typical SCC mixtures which has a possible influence on concrete behavior under load is the paste volume, in the same way as for shrinkage [NEV 00], since creep mechanisms are rooted at the level of the cement matrix. However, some researchers [CHO 03] are in disagreement with this on the basis of observations of similar creep deformation evolutions for two SCCs characterized by different paste volumes.

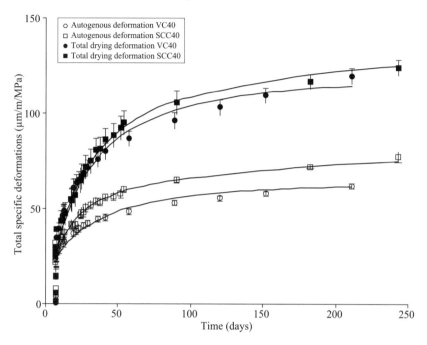

Figure 3.9. *Evolution of total deformation under load in autogenous and total drying modes for a SCC and a vibrated concrete in the same strength class, 40 MPa [VID 05]*

Other factors have not been analyzed nor have their effects been established (super-plasticizer type, grain shapes, etc.).

3.3.3.3. *Standard aspects*

Taking into account the importance of predicting the long term behavior of concretes subjected to sustained load, the majority of studies are compared with standard models. Predictions from Eurocode 2 (Figure 3.10) [PON 03, VID 05] or from the CEB version 90 model by [TUR 06] or version 99 [CEB 99] by [STA 07] appear to be adequate.

Landsberger *et al.* [LAN 07] established an experimental dataset for creep of conventional concretes and SCCs, which they compared with values calculated from different models (CEB, etc.). From this, they concluded that the models underestimate the specific creep of conventional concretes as well as those of SCCs, and that the differences are relatively low. On the other hand, the presence of mineral additions such as slag will strongly influence the creep size and kinetics, which is not taken into account in standard models.

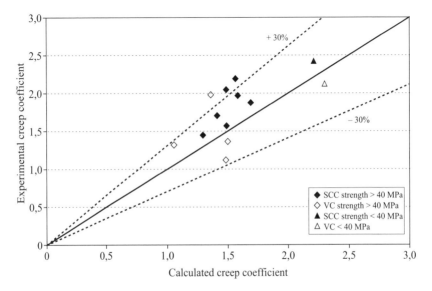

Figure 3.10. *Comparison between total creep coefficients of two SCCs and two vibrated concretes from strength classes lower and higher than 40 MPa, and comparison of their values with Eurocode 2 regulations [PON 03]*

3.4. Behavior of steel-concrete bonding

Having brought together the conclusions from research carried out on the intrinsic mechanical properties of SCCs, we now turn to steel-concrete bonding via the bond mechanism.

The bond between the two materials is a crucial characteristic with regard to the anchoring capacity of reinforcements in the concrete and the quality of stress transfer between the two materials which influences concrete cracking. In this section, we take apart the study of bonding and focus on the steel-SCC interface quality on these two aspects of its behavior.

3.4.1. *Anchorage capacity*

3.4.1.1. *Property studied and characterization*

The classic test used for evaluating the anchoring capacity of reinforcements in concrete is the pull-out test. This consists of applying an increasing pulling force on a steel reinforcement bar anchored in a concrete block. The relative slip between the steel and concrete is measured with a displacement probe.

From the bond stress τ versus slip curve as a function of the movement obtained, the ultimate bond stress τ_u is determined which corresponds to the maximum stress. The results can also be expressed as a function of the ratios $\tau_u/\sqrt{f_{c28}}$ (or τ_u/f_{c28} or τ_u/f_{t28} where f_{c28} and f_{t28} are the compressive and tensile strengths respectively of the concrete at 28 days) in order to normalize this value.

3.4.1.2. *Tendency and influencing factors*

The anchorage of ribbed reinforcement bars turns out to be the same, if not better, in the case of SCCs, according to the majority of researchers who have studied this behavior [CHA 03, VAL 09, CAT 09, CAS 06, ZHU 05, SON 99, SON 02].

Analysis of bond stress (sometimes normalized) versus slip curves shows a high gradient, but also and especially an ultimate, maximum stress (between 10 and 30% according to

[VAL 09]). However, we note that some studies, not numerous, have obtained disagreeing results with pull-out test strengths that are higher for conventional concretes [SCH 01, KON 01]. These can be explained by the wide variety of SCC and conventional concrete formulations used, as well as by the experimental conditions which are sometimes different in the various studies.

The quality of the steel-concrete interface is correlated to interface between the paste and the aggregates, known as the ITZ and described above. As in the case of the latter, the interface in SCCs is denser and less porous, which allows better rigidity and mechanical strength than for conventional concretes in the same strength class.

Steel-concrete bond mechanisms depend equally strongly on the top bar effect. This phenomenon arises from coupled effects of the loss of stability of the material when fresh, i.e. segregation, settlement and bleeding which favor the formation of defects at the interface and the presence of voids or localized water under horizontal reinforcement bars.

In order to check out the impact of the material type, SCC or conventional concrete, on this effect, researchers carried out a series of tests on samples taken from large size structures [SOY 03, KHA 98b, CAS 06]. The collected data showed that this phenomenon is much more dominant when the sample and its horizontal reinforcement bar are positioned high up in the formwork.

The reduction in bond stress noticed with conventional concrete between samples located at the top and the bottom of the formwork is strongly reduced for SCCs [VAL 09] (Figure 3.11) [CAS 06]. The combined use of super-plasticizer, thickening agents, high water reducer and a significant proportion of fine particles in the formulations allows an increased stability and homogeneity with reduced bleeding [KHA 97, KHA 98b].

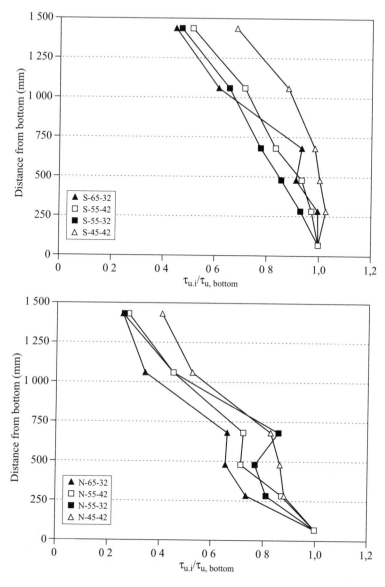

Figure 3.11. *Evolution of the ultimate bond stress ratio* $\tau_{u,i}$ *with its maximum value obtained at the base of the framework* $\tau_{u,\,bottom}$ *as a function of the position in height. Nomenclature: S (SCC), N (Conventional concrete), 55 (W/C=0.55), 65 (W/C=0.65), 32 (CEM 32.5), 42 (CEM 42.5), [VAL 09]*

[ZHU 05] showed through nano/micro-indentation analysis at the steel-concrete ITZ that the elastic modulus and compressive strength are effectively reduced by the fact that a locally increased W/C ratio develops in the zone below the reinforcement bar and this difference is less pronounced for SCCs.

When the top bar effect studies are carried out to compare SCCs and conventional concretes in higher strength classes, the differences become insignificant since the conventional concretes have a lower W/C ratio and become less susceptible to bleeding [VAL 09, CAS 06].

Although the use of "smooth" reinforcement bars is nowadays limited, the bond behavior of bars of this type has also been analyzed and the observations which come from these studies are rich in teachings. In effect, contrary to the majority of pull-out test data with ribbed reinforcements in SCCs, which give rise to a better anchorage in SCC, equivalent [CAS 06] or reversed results [DAO 03] have been found for smooth steels.

Studies on smooth reinforcements enable the physico-chemical adhesion behavior and friction between the two materials to be analyzed so that the studies on ribbed steel can analyze the blocking forces of concrete situated in the reinforcement ribs.

According to Daoud et al. [DAO 03], these forces are limited in SCCs because of the lower quantity of large aggregates. With the aid of video-microscope analysis, [CAS 06] showed that the debonding zone at the steel-concrete interface caused by the top bar effect was of a similar length, since the pull-out behaviors are similar.

On the other hand, this zone is wider. This provides an explanation with regard to the differences observed in the case of ribbed reinforcements. Too great a thickness of this comparative lack of bonding at the top of ribs provokes a noticeable reduction in the bond stress for conventional concretes.

3.4.1.3. *Building codes aspects*

Bond strength is estimated according to Eurocode 2 from concrete tensile strength. [ALM 08] noted that these predictions were reliable for conventional concrete as well as for SCCs.

Anchorage capacity is taken into account from the point of view of reinforced concrete structural design through the intermediary of an anchorage length. Valcuende *et al.* [VAL 09] recommended that the favorable effect of SCC in reducing its value by applying a minimizing coefficient as a function of the compressive strength be taken into account, and only for concretes in classes less than 50 MPa.

With regard to the top bar effect, some current codes take this phenomenon into consideration by involving correcting factors to increase the anchorage length of the reinforcements, when the height of the concrete under the bar goes over a limiting value [ACI 02, EUR 06]. Hence, accounting in this way can confine itself to conventional concretes, since SCCs have a reduced sensitivity to this effect.

3.4.2. *Transfer capacity of reinforcement tensile stress to concrete and cracking*

3.4.2.1. *Property studied and characterization*

A second mechanical aspect of steel-concrete interaction depends on bonding. This concerns stress transfer between the two materials. This mechanism is notable as the source of the stiffening effect of concrete under tension, known as tension stiffening. This occurs in structural elements which are subjected to flexing.

Tensile concrete zones located between flexural cracks behave as tensile members, with longitudinal steels which transfer tensile stress to the concrete. This results in a mechanical contribution of the tensile concrete which leads to relaxation in the reinforcement bars and a stiffening of flexural elements.

Furthermore, this property affects concrete cracking in terms of spacing and opening. When there is a good bond and an effective transfer of tensile forces between the two materials, the stress in the concrete can reach the level of its tensile strength, which results in the development of a new crack and further reduction in the spacing between cracks.

Transmission of stress between the two materials is assessed by using a test which consists of applying a tensile force to the two ends of a steel reinforcement bar running through the centre of a prismatic concrete element. The behavior of the concrete under tension is analyzed via measurements of the strains of its free surface and the development of cracks.

3.4.2.2. *Tendency and influencing factors*

Although this mechanical property is distinct from the anchorage capacity and has some importance for the design of reinforced concrete structures, studies are rare for SCCs.

A comparative study between the behavior of tension members using SCCs and conventional concretes [DAO 03] did not discern any difference, other than in terms of the crack network (distribution and opening) or a stress-strain relation along the element.

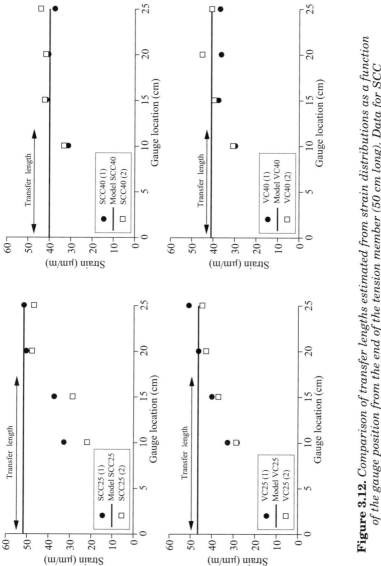

Figure 3.12. *Comparison of transfer lengths estimated from strain distributions as a function of the gauge position from the end of the tension member (50 cm long). Data for SCC and vibrated concretes from 25 MPa and 40 MPa classes*

From strains measured at the surface of a concrete and distributed along the tension member, the transfer length can be evaluated. This is defined as being the distance from the end of the tension member or from a crack, necessary for the reinforcement to transmit part of the tensile stress to the concrete in an optimum way.

Beyond this transfer length, the strain in the reinforcements stays constant. Relaxation in the steel, caused by the mechanical contribution of the concrete, is, accordingly, maximum. This means that, in that zone, the behavior of reinforced concrete is no longer influenced by the end of the tension member or the presence of cracks.

The values of these transfer lengths are *a priori* similar for the two concrete types, whether in the case of tension members [CAS 10] (Figure 3.12), or in structures with prestressed steels [ARB 03]. The top bar effect does not appear to have any impact on the transfer capacity of reinforcement tensile stress to concrete.

3.4.2.3. *Building code aspects*

Although authors have not touched upon building code aspects, we can deduce from their observations that taking into account concrete tension stiffening and calculating the opening of cracking in current codes can be generalized to SCCs.

3.5. Bibliography

[ACI 02] ACI 318, Building code requirements for structural concrete (ACI 318-02) and commentary (ACI 318R-02), ACI Committee 318 Structural Building Code, American Concrete Institute, 2002.

[ALM 08] ALMEIDA FILHO F.M., EL DEBS M.K., EL DEBS A.L., "Bond-slip behavior of self-compacting concrete and vibrated concrete using pull-out and beam tests", *Materials and Structures*, vol. 41, p. 1073-1089, 2008.

[ARB 03] ARBELAEZ JARAMILLO C.A., RIGUEIRA VICTOR J.W., MARTI VARGAS J.R., SERNA ROS P., PINTO BARBOSA M., "Bond characteristic of pre-stressed strands in self-compacting concrete", *3rd International RILEM Symposium Self-Compacting Concrete*, p. 684-691, Reykjavik, Iceland, 2003.

[ASS 03] ASSIE S., ESCADEILLAS G., MARCHESE G., "Durability of self-compacting concrete: a different behaviour compared with vibrated concrete?", *3rd International RILEM Symposium Self-Compacting Concrete*, p. 655-662, Reykjavik, Iceland, 2003.

[AST 05] ASTM C 1581 – 04, Standard test method for determining age of cracking and induced tensile stress characteristics of mortar and concrete under restrained shrinkage, ASTM, 2005.

[BON 00] BONAVETTI V., DONZA H., RAHHAL V., IRASSAR E., "Influence of initial curing on the properties of concrete containing limestone blended cement", *Cement and Concrete Research*, vol. 30, no. 5, p. 703-708, 2000.

[BOS 03] BOSILJKOV V.B., "SCC mixes with poorly graded aggregate and high volume of limestone filler", *Cement and Concrete Research*, vol. 33, no. 9, p. 1279-1286, 2003.

[BPE 99] BPEL 91, Règles techniques de conception et de calcul des ouvrages et constructions en béton précontraint suivant la méthode des états limites, revised 99, Issue 62, Title I, section II, April 1999.

[CAS 06] CASTEL A., VIDAL T., VIRIYAMETANONT K., FRANÇOIS R., "Effect of reinforcing bar orientation and location on bond with self-consolidating concrete", *ACI Materials Journal*, vol. 103, no. 4, p. 559-567, 2006.

[CAS 10] CASTEL A., VIDAL T., FRANÇOIS R., "Bond and cracking properties of self-consolidating concrete", *Construction and Building Materials*, vol. 24, no. 7, p. 1551-1561, 2010.

[CAT 09] CATTANEO S., ROSATI G., "Bond between steel and self-consolidating concrete: experiments and modeling", *ACI Structural Journal*, vol. 106, no. 4, p. 540-550, 2009.

[CEB 90] Comité Euro-International du Béton, 1991, CEB-FIB Model Code 1990, Design Code, Thomas Telford Publications, Lausanne, Switzerland, 1990.

[CEB 99] CEB-FIP, "Structural concrete volume 1", *Bulletin fib*, no. 1, 1999.

[CHA 03] CHAN Y.W., CHEN Y.S., LIU Y.S., "Development of bond strength of reinforcement steel in self-consolidating concrete", *ACI Structural Journal*, vol. 100, no. 4, p. 490-498, 2003.

[CHO 03] CHOPIN D., FRANCY O., LEBOURGEOIS S., ROUGEAU P., "Creep and shrinkage of heated-cured self-compacting concrete (SCC)", *3rd International RILEM Symposium Self-Compacting Concrete*, p. 672-683, Reykjavik, Iceland, 2003.

[DAO 03] DAOUD A., LORRAIN M., LABORDERIE C., "Anchorage and cracking behaviour of self-compacting concrete", *3rd International RILEM Symposium Self-Compacting Concrete*, p. 692-702, Reykjavik, Iceland, 2003.

[DIN 07] DINAKAR P., BABU K.G., SANTHANAM M., "Mechanical properties of high volume of ash self-compacting concretes", *5th International RILEM Symposium Self-Compacting Concrete*, p. 651-658, Ghent, Belgium, 2007.

[DOM 07] DOMONE P.L., "A review of the hardened mechanical properties of self-compacting concrete", *Cement and Concrete Composites*, vol. 29, p. 1-12, 2007.

[EUR 06] Eurocode 2 EN1992-1-1 Calcul des structures en béton armé, partie 1.1., règles générales et règles pour les bâtiments, 2006.

[FEL 07] FELEKOGU B., TURKEL S., BARADEN B., "Effect of water/cement ratio on the fresh and hardened properties of self-compacting concrete", *Building and Environment*, vol. 42, p. 1795-1802, 2007.

[FER 07] FERNANDEZ-GOMEZ J., LANDSBERGER G.A., "Evaluation of shrinkage prediction models for self-consolidating concrete", *ACI Materials Journal*, vol. 104, no. 5, p. 464-473, 2007.

[HEI 03] HEIRMAN G., VANDEWALLE L., "The influence of fillers on the properties of self-compacting concrete in fresh and hardened state", *3rd International RILEM Symposium Self-Compacting Concrete*, p. 606-618, Reykjavik, Iceland, 2003.

[HEI 08] HEIRMAN G., VANDEWALLE L., VAN GEMERT D., BOEL V., AUDENAERT K., DE SCHUTTER G., DESMET B., VANTOMME J., "Time dependent deformations of limestone powder self-compacting concrete", *Engineering Structures*, vol. 30, no. 10, p. 2945-2956, 2008.

[HWA 10] HWANG S.D., KHAYAT K.H., "Effect of mix design on restrained shrinkage, of self-consolidating concrete", *Materials and Structures*, vol. 43, p. 367-380, 2010.

[JAC 99] JACOBS F., HUNKELER F., "Design of self-compacting concrete for durable concrete structures", *First International RILEM Symposium on Self-Compacting Concrete*, p. 397-410, Rilem Publications, Stockholm, Sweden, 1999.

[KHA 97] KHAYAT K.H., MANAI K., TRUDEL A., "*In situ* mechanical properties of wall elements cast using self-consolidating concrete", *ACI Materials Journal*, vol. 94, no. 6, p. 491-500, 1997.

[KHA 98a] KHAYAT K.H., "Viscosity-enhancing admixtures for cement-based materials – an overview", *Cement and Concrete Composites*, vol. 20, p. 171-188, 1998.

[KHA 98b] KHAYAT K.H., "Use of viscosity-modifying admixture to reduce top-bar effect of anchored bars cast with fluid concrete", *ACI Materials Journal*, vol. 95, no. 2, p. 158-167, 1998.

[KLU 03] KLUG Y., HOLSCHEMACHER K., "Comparison of the hardened properties of self-compacting and normal vibrated concrete", *3rd International RILEM Symposium Self-Compacting Concrete*, p. 596-605, Reykjavik, Iceland, 2003.

[KON 01] KONIG G., HOLSCHEMACHER K., DEHN F., WEISSE D., "Self-compacting concrete time development of material properties and bond behavior", *Proceedings of Second International Symposium on Self-compacting Concrete*, p. 507-516, Tokyo, Japan, 2001.

[KON 03] KONIG G., HOLSCHEMACHER K., DEHN F., WEISSE D., "Bond of reinforcement in self-compacting concrete (SCC) under monotonic and cycling loading", *3rd International RILEM Symposium Self-Compacting Concrete*, p. 939-947, Reykjavik, Iceland, 2003.

[LAN 07] LANDSBERGER G.A., FERNANDEZ-GOMEZ J., "Evaluation of creep prediction models for self consolidating concrete", *5th International RILEM Symposium Self-Compacting Concrete*, p. 605-610, Ghent, Belgium, 2007.

[LOS 09] LOSER R., LEEMANN A., "Shrinkage and restrained shrinkage cracking of self-compacting concrete compared to conventionally vibrated concrete", *Materials and Structures*, vol. 42, p. 71-82, 2009.

[MAZ 05] MAZZOTTI C., SAVOIA M., CECCOLI C., "A comparison between long term properties of self-compacting concretes and normal vibrated concretes with same strength", *CONCREEP7 Creep, Shrinkage and Durability of Concrete and Concrete Structures*, p. 523-528, ISTE, London, 2005.

[MAZ 07] MAZZOTTI C., SAVOIA M., CECCOLI C., "Comparison between long term properties of self-compacting concretes with different strengths", *5th International RILEM Symposium Self-Compacting Concrete*, p. 599-604, Ghent, Belgium, 2007.

[MNA 08] MNAHONCAKOVA E., PAVLIKOVA M., GRZESZCZYK S., ROVNANIKOVA P., CERNY R., "Hydric, Thermal and mechanical properties of self-compacting concrete containing different fillers", *Construction and Building Materials*, vol. 22, no. 7, p. 1594-1600, 2008.

[NEV 00] NEVILLE A.M., *Propriétés des bétons*, Eyrolles, Paris, 2000.

[OGA 95] OGAWA A., SAKATA K., TANAKA S., "A study on reducing shrinkage of high-flowable concrete", *2nd International Symposium of CANMET/ACI*, p. 55-72, Las Vegas, USA, 1995.

[PAR 07] PARRA C., VALCUENDE M., BENLLOCH J., "Mechanical properties of self-compacting concretes", *5th International RILEM Symposium Self-Compacting Concrete*, p. 599-604, Ghent, Belgium, 2007.

[PER 99] PERA J., HUSSON S., GUILHOT B. "Influence of finely ground limestone on cement hydration", *Cement and Concrete Composites*, vol. 21, no. 2, p. 99-105, 1999.

[PER 01] PERSSON B., "A comparison between mechanical properties of self-compacting concrete and the corresponding properties of normal concrete", *Cement and Concrete Research*, vol. 31, no. 2, p. 193-198, 2001.

[PER 05] PERSSON B., "Creep of self-compacting", *CONCREEP7 Creep, Shrinkage and Durability of Concrete and Concrete Structures*, p. 535-540, Nantes, 2005.

[PON 03] PONS G., PROUST E., ASSIE S., "Creep and shrinkage of self-compacting concrete: a different behaviour compared with vibrated concrete?", *3rd International RILEM Symposium Self-Compacting Concrete*, p. 645-654, Reykjavik, Iceland, 2003.

[ROL 97] ROLS S., AMBROISE J., PERA J., "Development of an admixture for self-leveling concrete", *Proceedings of the 5th CANMET/ACI International Conference on Superplastifiant and other Chemical Admixtures in Concrete*, p. 493-509, Rome, Italy, 1997.

[RON 02] RONCERO J., GETTU R., "Influence of super plasticizers on the microstructures of hydrated cement paste and creep behaviour of cement mortar", *Cemento-Hormigon*, vol. 832, p. 12-28, 2002.

[ROZ 07] ROZIÈRE E., GRANGER S., TURCRY P., LOUKILI A., "Influence of paste volume on shrinkage cracking and fracture properties of self-compacting concrete", *Cement and Concrete Composites*, vol. 29, no. 8, p. 26-636, 2007.

[RUL 00] RULE BOOK, *Béton armé aux états limites – réglementation française de conception, calcul et construction des ouvrages en béton armé par la méthode des états limites*, Eyrolles, 2000.

[SCH 01] SCHIESSL A., ZILCH K., "The effects of the modified composition of SCC on shear and bond behavior", *Proceedings of 2nd International Symposium on Self-Compacting Concrete*, p. 501-506, Tokyo, Japan, 2001.

[SCH 07] SCHINDLER A.K., BARNES R.W., ROBERTS J.B., RODRIGUEZ S., "Properties of self-consolidating concrete for prestressed members", *ACI Materials Journal*, vol. 104, no. 1, p. 53-61, 2007.

[SON 99] SONEBI M., BARTOS P.J.M., "Hardened SCC and its bond with reinforcements", *Proceeding of RILEM International Symposium on SCC*, p. 275-290, Stockholm, Sweden, 1999.

[SON 02] SONEBI M., BARTOS P.J.M., "'Bond behavior and pull-off test of self compacting concrete', bond in concrete, from research to standards", *Proceedings of the 3rd International Symposium*, p. 511-519, Budapest, Hungary, 2002.

[SOY 03] SOYLEV T.A., FRANCOIS R., "Quality of steel-concrete interface and corrosion of steel", *Cement and Concrete Research*, vol. 33, p. 1407-1414, 2003.

[STA 06] STAQUET S., BOULAY C., D'ALOIA L., TOUTLEMONDE F., "Autogenous shrinkage of a self-compacting VHPC in isothermal and realistic temperature conditions", *2nd International Symposium on Advances in Concrete through Science and Engineering*, Québec, Canada, 2006.

[STA 07] STAQUET S., BOULAY C., D'ALOIA L., LE ROY R., ESPION B., TOUTLEMONDE F., "Creep and shrinkage in prebent composite beams for innovative railway bridges in France – Part I: modeling", *American Concrete Institute Fall Convention*, American Concrete Institute Separate Publication, *Structural Implications of Shrinkage and Creep of Concrete, SP-246*, 2007, Paper 4, Puerto Rico, p. 53-70, 2007.

[TAN 93] TANAKA K., SATO K., WATANABE S., ARINA I., SUENAGA K., "Development and utilization of high performance concrete for the construction of the Akashi Kaikyo Bridge", *International Symposium on High Performance Concrete in Science Environments*, p. 25-51, SP140 ACI, Detroit, USA, 1993.

[TUR 06] TURCRY P., LOUKILI A., HAIDAR K., PIJAUDIER-CABOT G., BELARBI A., "Cracking tendency of self-compacting concrete subjected to restrained shrinkage: experimental study and modeling", *Journal of Materials in Civil Engineering ASCE*, vol. 18, no. 1 p. 46-54, 2006.

[VAL 09] VALCUENDE M., PARRA C., "Bond behaviour of reinforcement in self-compacting concretes", *Construction and Building Materials*, vol. 23, p. 162-170, 2009.

[VID 05] VIDAL T., ASSIE S., PONS G., "Creep and shrinkage of self-compacting concrete and comparative study with model code", *CONCREEP7 Creep, Shrinkage and Durability of Concrete and Concrete Structures*, p. 541-546, Nantes, 2005.

[VIE 03] VIEIRA M., BETTENCOURT A., "Deformability of hardened SCC", *3rd International RILEM Symposium Self-Compacting Concrete*, p. 637-644, Reykjavik, Iceland, 2003.

[XIE 05] XIE Y., LI Y., LONG G., "Influence of aggregate on properties of self consolidating concrete", *RILEM Proceedings of the First International Symposium on Design, Performance and Use of Self-Consolidating Concrete*, p. 161-171, Changsha, China, 2005.

[ZHU 05] ZHU W., BARTOS P.J.M., "Microstructure and properties of interfacial transition zone in SCC", *RILEM Proceedings of the First International Symposium on Design, Performance and Use of Self-Consolidating Concrete*, p. 319-327, Changsha, China, 2005.

Chapter 4

Durability of Self-Compacting Concrete

4.1. Introduction

Self-compacting concrete (SCC) is subjected to the same durability requirements as conventional concrete. The specifics of their formulations legitimately lead to questions regarding their resistance to environmental effects, in comparison with those of conventional concrete. However, SCC compositions are as varied as conventional concrete compositions and we can easily see that an *a priori* response to this question is not possible. The problem lies in defining and evaluating SCC durability.

The simplest way of responding to this problem is to consider prescriptive requirements, in the sense of standards [NFE 04]. The characteristic compressive strength is used as the principal durability indicator and the standard prescribes, as a function of strength and environmental conditions, minimal requirements in terms of equivalent binder (eq. binder) content and effective water/eq. binder

Chaper written by Emmanuel ROZIÈRE and Abdelhafid KHELIDJ.

ratios. SCC compositions generally satisfy these requirements. But this approach does not directly take into account composition specifics such as the paste volume, the Gravel/Sand ratio, the mineral admixture content, and the type of mineral admixtures – the coefficient for taking into account the mineral admixtures in terms of the eq. binder is based on its contribution to the strength, not durability, of the concrete.

The definition of exposure classes introduced in Standard EN 206-1 allows concrete degradation in a given environment to be taken into account. This therefore opens up a way of evaluating the durability performance of a concrete. This evaluation is not of the composition, but of the potential resistance of concrete to the degradation to which it is exposed. This appears, therefore, to be well adapted to the problem of self-compacting concrete durability. Three approaches are possible:

1. The compressive strength, porosity, gas permeability, and diffusion coefficients constitute *performance* indicators, since they characterize the material other than by its composition – they are often described as *general* since these indicators are not specific to a degradation mode. Since they are also the input data for some models, in a reversed approach, the yield points are therefore proposed as a function of the desired service life of the structure [BAR 04].

2. The comparative approach, which entails subjecting several concrete mixtures to the same ageing test, is set out in the Standard NF EN 206-1 [NFE 04] as the *Equivalent Performance Concept*. Subject to the representativeness of the test in relation to real degradation mechanisms, this approach enables a comparative answer to the durability question. It has often been applied to SCC, for comparisons with other SCC or conventional concrete mixtures.

3. Field data can give a foundation for good concrete behavior in a given environment. This approach is seldom

used, because of the evolution over time of component characteristics.

This chapter approaches the question of durability from two angles, with regard to the two approaches based on performance indicators. In sections 4.1 and 4.2, general indicators are presented such as the porosity, the gas permeability and the chloride ion diffusion coefficient, and information on their variability. Section 4.3 is concerned with degradation mechanisms and specific ageing tests.

Comparisons show that, in return for some care with the formulations, the potential durability of SCC is generally equivalent to that of conventional concrete [AUD 03, TUR 04, ASS 04]. But this chapter is, above all, intended to provide the core knowledge – common to all concrete types – necessary to take into account durability in formulating SCC, and to provide a few tools for mastering durability and data for optimizing compositions.

4.2. Properties and parameters that influence durability

4.2.1. *Mechanical strength*

Without being specific to any degradation mechanism – two concrete mixtures of the same strength can show radically different behavior – the compressive strength is a performance indicator which is of interest in several aspects. Firstly, as the principal indicator in the standard, it is the first criteria for comparisons of SCC mixtures with conventional concrete mixtures.

Secondly, its experimental procedure has been standardized for a long time and an abundant dataset is therefore available, as a function of different composition parameters (mixture proportions, cement types and chemical additives, etc.). Finally, it is a sensitive indicator: small variations in composition lead to significant variations in the property.

The effect of the water/cement ratio (W/C) or the effective water/cement ratio (W_{eff}/C) (to account for the water which is potentially absorbed by the aggregates and therefore not available for hydrating the binder) is known. Simple models (Féret and Bolomey equations) enable optimization of this ratio as a function of the target strength.

Table 4.1 shows the compressive strength at 28 days against the W_{eff}/C ratio for Assié concrete mixtures [ASS 04] – self-compacting concrete (SCC) and conventional concrete (CC).

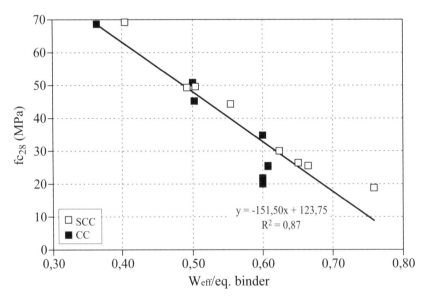

Figure 4.1. *Change in compressive strength as a function of the W_{eff}/eq. binder ratio, using data from Assié [ASS 04]*

Composition (kg/m³)	Concrete mixtures							
	Series	SCC 15	SCC 20	CC20	SCC 40	CC40	SCC 60	CC60
		CEM II/A-LL 32, 5 R			CEM I 52, 5 N			
Cement		315	315	315	350	350	450	450
Limestone filler		150	150	-	140	-	70	-
Eq. binder		315	315	315	379. 2	350	467. 5	450
Effective water	I		196. 5	191. 4	210. 1	210. 0		
	II	238. 9	209. 5	189. 0	186. 5	175. 8	189. 0	163. 8
	III		205. 0	189. 0	191. 0	175. 0		
W_{eff}/cement	I		0. 62	0. 61	0. 60	0. 60		
	II	0. 76	0. 67	0. 60	0. 53	0. 50	0. 42	0. 36
	III		0. 65	0. 60	0. 55	0. 50		
W_{eff}/eq. binder	I		0. 62	0. 61	0. 55	0. 60		
	II	0. 76	0. 67	0. 60	0. 49	0. 50	0. 40	0. 36
	III		0. 65	0. 60	0. 50	0. 50		
Paste volume	I der		385. 0	320. 0	388. 7	333. 4		
	II	423. 8	382. 7	310. 4	366. 8	307. 0	376. 8	318. 0
	III		379. 8	316. 6	374. 4	304. 8		
Trapped air (%)	I		3. 4	7. 8	1. 8	3. 6		
	II	-	2. 1	5. 2	1. 3	4. 1	1. 6	1. 6
	III		2. 1	7. 6	2. 2	4. 0		
Gravel/sand		0. 86	0. 86	0. 86	0. 89	0. 89	0. 90	1. 45
fc_{28} (MPa)	I		30. 0	25. 4	44. 3	34. 8		
	II	18. 8	25. 5	20. 0	49. 3	45. 2	69. 2	68. 6
	III		26. 4	21. 8	49. 6	50. 8		
Porosity (%)	I		14. 9	14. 9	13. 7	14. 7		
	II	18. 1	16. 2	15. 2	12. 5	11. 2	10. 5	8. 9
	III		15. 2	14. 2	13. 6	11. 7		

Table 4.1. *Compositions of SCC and conventional concrete mixtures [ASS 04]*

SCC seems to show higher strength than conventional concrete. SCC compositions actually incorporate a significant quantity of limestone filler (between 70 and 150 kg/m³ here). Hence the concrete mixtures have the strength of conventional concrete mixtures with lower W_{eff}/C ratios, which shows the contribution of filler as much as it shows the binder. The concept of eq. binder (NF EN 206-1) takes into account this contribution (Figure 4.1). However, the accounting coefficient for the filler of 0.25 seems to underestimate this effect. We can also note the relatively low strength of some conventional concrete mixtures, formulated for a slump of around 100 mm, and which have high entrapped air contents. Generally, the entrapped air content of conventional concrete is higher than that of SCC, which can explain these results. We can therefore assume that the SCC fluidity when fresh, which is linked to a better defloculation of fine particles in the cement, ensures better compactness.

The influence of the paste volume (defined as the cement, mineral admixtures, effective water and air) is less commonly addressed. Furthermore, variations are generally lower for conventional concrete than for SCC mixtures, and the variation ranges are also different: between 240 and 300 L/m³ for conventional concrete and between 330 and 420 L/m³ for SCC. The general tendency for SCC is a reduction in compressive and tensile strength with increases in paste volume [ROZ 07]. At 28 days, when the paste volume changes from 291 to 457 L/m³, an increase of 57%, the compressive strength reduces from 47.1 to 41.7 MPa, a reduction of 12% (Figure 4.2). Variation in the paste volume therefore seems to have a real but limited effect on the compressive strength of self-compacting concrete. The tensile strength reduces with paste volume in the same ratio as the compressive strength: it is reduced by around 13% in the same range. This observation is useful when the paste composition and workability are constant. If the increase in paste volume is due to an increase in the effective water, the loss in strength is much more noticeable (Figure 4.2).

This tendency has already been observed experimentally in conventional concrete [STO 79] and in SCC [TUR 04, KOL 04]. De Larrard's model [LAR 99] enables this result to be found via the maximum paste thickness, defined as the space which separates two large grains of size D_{max} (initially in contact) after the injection of the binder matrix in the dry grain piling, optimized from the point of view of compactness.

De Larrard also noted that the effect of the aggregate volume, which is not uniform, can be masked by the increase in entrapped air when workability reduces.

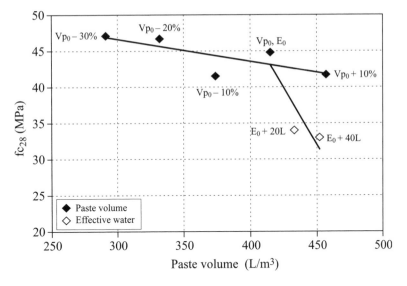

Figure 4.2. *Compressive strength as a function of paste volume [ROZ 07]*

4.2.2. *Porosity and properties of the porous network*

Porosity is an indicator of concrete quality since it enables detection of production variability. It is also a durability indicator since it is involved in most degradation mechanisms. It is therefore input data for the mathematical modeling of degradations. From these models, porosity yield points have been proposed as targets for structure lifespans [BAR 04]. However, the properties of porous network are at least as influential as the total porosity: connectivity, tortuosity, pore size distribution.

4.2.2.1. *Composition influences*

The W_{eff}/eq. binder ratio is the dominant influence on porosity (Table 4.1). Extra porosity in SCC is therefore explained by the differences in the effective water, the constant in the study being the cement content [ASS 04].

In hypothesizing that all of the water-accessible porosity in the concrete is the paste porosity, the porosity related to the paste volume can be calculated from equation [4.1].

$$Porosity_{Concrete} = Porosity_{Paste}.Paste\ Volume \qquad [4.1]$$

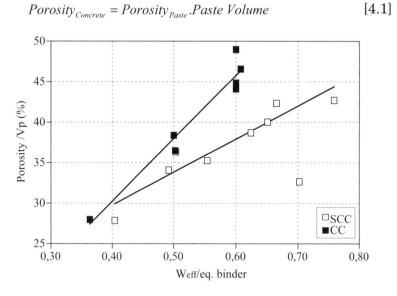

Figure 4.3. *Changes in the paste porosity as a function of the W_{eff}/eq. binder ratio, using data from Assié [ASS 04]*

If the influence of the W_{eff}/eq. binder ratio is always well observed, it also appears that the paste porosity is low for SCC (Figure 4.3). This may be due to the lower entrapped air contents of SCC mixtures, with respect to their more fluid consistency.

4.2.2.2. *Properties of the porous network*

Two concretes of the same overall porosity can show difference behaviors and therefore different durability. As seen above, these differences can arise from the paste volume and the effective porosity of the paste. The differences can be explained by the properties of the connected porous network, i.e. the porous volume accessible to aggressive aqueous or gaseous reagents, and the change in properties over time, since exposure to aggressive reagents is often early and long-term behavior is influenced by the early age properties.

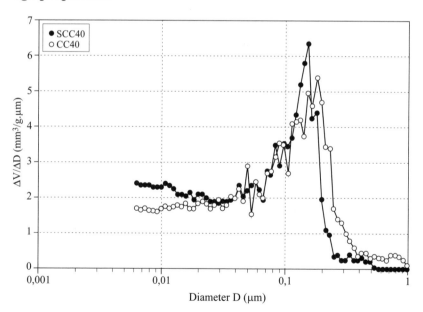

Figure 4.4. *Porosity distribution as a function of pore diameter. Compositions SCC40 and CC40 [TUR 04]*

In spite of some limits, mercury intrusion porosimetry is an interesting tool for studying the evolution of pore size distribution over time. Figures 4.4 and 4.5 show porosity distribution as a function of the pore diameter at 2 days. This illustration has the advantage of giving main pore sizes. There is no noticeable distribution particular to conventional concrete or SCC. At 2 days, SCC40 has smaller pores than CC40. On the other hand, SCCF5 has wider pores than CC5. Each concrete has only one peak, which corresponds to capillary pores.

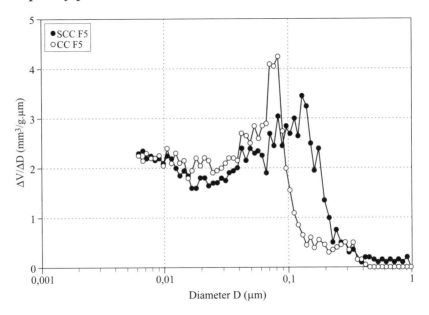

Figure 4.5. *Porosity distribution as a function of pore diameter. Compositions SCC F5 and CC F5 [TUR 04]*

Over time, the pores fill with hydrates; the observed peak moves towards smaller pore sizes and reduces in height. In more compact concrete, porosity becomes very low and the peak may even tend to disappear. In Table 4.2, the difference between pore diameter for the SCC40 and CC40 peaks is seen to be almost identical at 2 and 28 days (in the order of

15 nm). On the other hand, the difference between the SCCF5 and CCF5 peaks reduces over time (from 50 to 1 nm).

	SCC40	CC40	SCCF5	CCF5
Peak at 2 days (nm)	152	167	131	83
Peak at 28 days (nm)	58	72	56	57

Table 4.2. *Principal porosity peaks for SCC and conventional concrete [TUR 04]*

4.2.3. *Absorption*

If concrete is partially saturated or subjected to cycles of humidity and dry conditions, as is the case for parts of structures which are exposed to tides, sea spray and de-icing salts, salt is likely to migrate in the interstitial liquid phase by convection [FRA 01].

Water absorption is due to surface tension in the capillaries. This depends not only on the porous network, but also on the degree of saturation of the concrete [BAS 01]. Water absorption in dry concrete is represented by two principal parameters:

– the quantity of water required to completely saturate the material (the porosity);

– the absorption rate by capillary action or absorptivity (denoted S).

The behavior of concrete subjected to saturation tests can hence be described by equation [4.2]:

$$A(t) = C + St^{1/2} \qquad\qquad [4.2]$$

where $A(t)$ is the capillary absorption coefficient for a given test period, in kg per m^2 of surface exposed to saturation.

The initial water absorption coefficient (the mass of water absorbed after one hour of suction) was measured for concrete subjected to an accelerated carbonation test [DES 04]. The carbonation depth increases with this coefficient, regardless of the concrete type. However, the absorption coefficients after the next hour can be very different. In effect, the diffusion of CO_2 in the concrete depends on the internal hygrometry, amongst other factors, which cannot have any influence on the water absorption coefficient at one hour since this measurement only characterizes the widest capillaries. This water absorption coefficient therefore only gives information on carbonation for concrete which has internal degree of saturation of concretes that are close by, and this internal degree of saturation of concrete is linked to the treatment mode before exposure to carbonation. It is nevertheless interesting to consider the influence of this indicator, since it is easily accessed; it is the object of a normalized operating mode (annex G of NF EN 13369), and it replaces the minimum cement content criteria in table NA.F.2 (specification for ready-to-use concrete) in Standard EN 206-1 [NFE 04].

4.3. Transport phenomena

4.3.1. *Permeability*

Permeability is the transport mode for matter linked to a pressure gradient. With the exception of walls for confining nuclear reactors or waste, there are few concrete structures subjected to gas pressure gradients. Permeability is, however, an indicator that is often used to evaluate the potential durability of concrete [BAR 04], since it characterizes the extent and coarseness of the interconnected porous network.

4.3.1.1. *Measuring gas permeability: determining apparent and intrinsic permeability*

Permeability can *a priori* be determined for any fluid (water or gas). However, because of the interactions that are possible between the cement matrix and water, a flow of inert gas, such as oxygen or nitrogen, is often preferred. For a gas, permeability depends on the test pressure (equation [4.3]) and the degree of saturation of concrete. Intrinsic permeability (equation [4.4]), when determined according to the Klinkenberg approach, results in a property which is independent of pressure. It is often determined for the dry material, which is not always representative with regard to real degradation conditions but allows a characteristic to be obtained which is independent of the degree of saturation of concrete.

For each sample and pressure, the apparent permeability K_a can by calculated using equation [4.3].

$$K_a = \frac{Q_1 2\mu L P_{atm}}{S\left(P_1^2 - P_{atm}^2\right)}$$ [4.3]

– Q_1: input flow (m³/s);

– μ: dynamic viscosity of nitrogen (Pa.s);

– P_1: injection pressure (Pa);

– P_{atm}: atmospheric pressure (Pa);

– L: sample thickness (m);

– S: sample area (m²).

From measuring the apparent permeability at several pressures, the intrinsic permeability K_v can be deduced using the Klinkenberg approach [KLI 41] via equation [4.4].

$$K_a = K_v(1 + \frac{\beta}{P_m})$$ [4.4]

– P_m: average pressure, equal to $(P_1 + P_{atm})/2$;

– β: Klinkenberg coefficient.

The coefficient β is a function of the coarseness of the porous network and the type of gas. The intrinsic permeability K_v is the limiting value of the apparent permeability when the average pressure of the fluid tends to infinity, i.e. the gas tends towards a condensed phase (liquid). K_v is determined from a linear regression of the apparent permeability measurements taken at different injection pressures according to the inverse of the average pressure [PIC 01].

4.3.1.2. *Permeability and potential durability*

Relations have been proposed to link permeability and potential durability, in particular for the case of carbonation and the risk of corrosion of reinforcement bars [RIL 99, BAS 01, BAR 04].

However, a clear correlation has not been observed between the carbonation rate and gas permeability of concrete, even within a single strength class [ASS 04, ROZ 09]. Since transport of chemical species is actually due to differences in concentration rather than pressure, it is more similar to diffusion. But permeability tests can provide interesting information with regard to the ease with which O_2 in the air can get to reinforcements that are likely to corrode. Intrinsic permeability can be a more relevant indicator than apparent permeability, insofar as it characterizes the material independently of the measurement pressure. However, because of the changes in pore types as a result of carbonation, gas permeability measured for sound concrete cannot be considered as a relevant indicator from this perspective. On the other hand, the test clearly demonstrates the properties of the sample surface layer, which is the first to be exposed to carbonation.

This explains why gas permeability is proposed for use as a carbonation indicator.

4.3.1.3. *Permeability of SCC and conventional concrete*

Permeability is a function of pore size, which depends strongly on the age of concrete, the W_{eff}/C ratio and the binder type. The higher the W_{eff}/C ratio, the higher the capillary porosity and permeability increases [PER 99]. This influence is also true for SCC.

The higher paste volumes in SCC mixtures might suggest that the permeability is increased as a consequence. However, measurements carried out on SCC mixtures with a constant paste composition, but variable proportions (Figure 4.6), do not show significant differences: the apparent permeability values measured with a pressure of 4 bar are low and of the same order of magnitude.

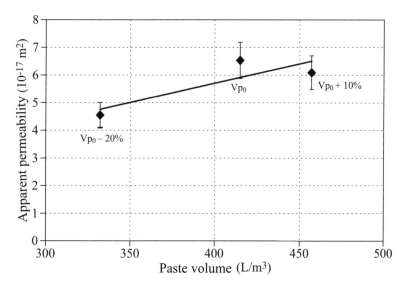

Figure 4.6. *Apparent permeability measured under a pressure of 4 bars (GeM tests)*

The dominant effect seems to be that of additions such as limestone filler and fly ash, of which high proportions are a characteristic of SCCs. Several studies [TRA 99, ZHU 01, DES 03] have shown that SCC permeability is lower than that of corresponding conventional concrete mixtures, and that this reduction is due to the use of mineral admixtures. By showing the evolution of intrinsic permeability as a function of the W_{eff}/eq. binder ratio, using data from Assié (Figure 4.7), it is clear that SCC permeability is significantly lower, for a given W_{eff}/eq. binder ratio. The accounting coefficient for limestone filler, in calculating the eq. binder, therefore underestimates the contribution of filler to the reduction in permeability.

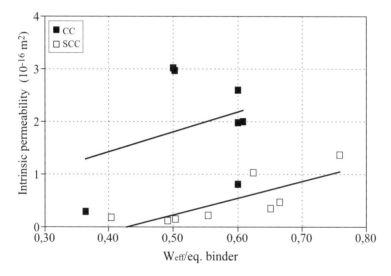

Figure 4.7. *Evolution in intrinsic permeability as a function of W_{eff}/eq. binder, after data from Assié [ASS 04]*

4.3.2. *Diffusion*

4.3.2.1. *Transport by diffusion*

Diffusion transport of a species is linked to its concentration gradient. In an ideal solution, i.e. one which is infinitely diluted, the electrochemical interactions can be ignored which allows the diffusive ion flux to be expressed as:

$$\vec{J} = -D\overrightarrow{grad}(c) \qquad [4.5]$$

where c is the concentration of the species under consideration and D is the diffusion coefficient of this ion in solution.

In an elementary volume, in which the concentration difference between two points is equal to the quantity of the material exchanged by diffusion, the equation of conservation of the material is written (in differential notation):

$$\frac{\partial c}{\partial t} = -div(\vec{J}) = div(D\overrightarrow{grad}(c)) \qquad [4.6]$$

The diffusion coefficient D for the ion in solution is a size which includes parameters that are particularly linked to ion mobility and therefore to ion diameter and ionic forces in the solution. If the porous region is considered to be the representative elementary volume, other parameters such as the network complexity or connectivity also influence ionic movements at the heart of the porous region.

4.3.2.2. Diffusion of chloride ions in concrete

Only a fraction of the chloride ions penetrating the concrete stay in solution. In effect, chlorides can react with hydrates in the cement paste, and a non-negligible fraction of chloride ions bond to the cement matrix. The following are distinguished:

– free chlorides ClF, which are present in ionic form in solution;

– bound chlorides ClB, adsorbed on CSH or chemically bonded as calcium chloroaluminates (particularly as calcium

hydrate   mono-chloroaluminate)   or   Friedel   salt,
$C_3A.CaCl_2.10H_2O$).

Desorption may occur and chloride ions can return to their ionic form [FRA 01].

The volumetric content of total chlorides (ClT), denoted m_{ClT}, is therefore written:

$$m_{ClT} = m_{ClF} + m_{ClB}$$ [4.7]

Applying Fick's first law to (free) chloride ions in concrete:

$$\overrightarrow{J_{ClF}} = -D_{eff}\,\overrightarrow{grad}(c_{ClF})$$ [4.8]

where D_{eff} is the effective diffusion coefficient for chloride ions, which takes into account porous network characteristics.

From equation [4.7], the chloride ions transfer equation is written as follows, commonly known as Fick's second law:

$$\frac{\partial m_{ClT}}{\partial t} = div\left[D_{eff}\,\overrightarrow{grad}(c_{ClF})\right]$$ [4.9]

If the diffusion coefficient D_{eff} is assumed to be constant, i.e. it is independent of the concentration of free chloride ions, we find:

$$\frac{\partial c_{ClF}}{\partial t} = D_{app}\frac{\partial^2 c_{ClF}}{\partial x^2}$$ [4.10]

with:

$$D_{app} = \frac{D_{eff}}{p + \rho\dfrac{\partial M_{ClB}}{\partial c_{ClF}}}$$

An apparent diffusion coefficient has been defined, which depends on the effective diffusion coefficient D_{eff}, the porosity p, and the chloride ion bonding to the cement matrix as a function of the concentration.

The paste bonding capacity for chloride ions depends on the temperature and the binder type. Several equations for the function $M_{ClB}=f(c_{ClF})$ can be found in the literature.

Diffusion of chloride ions in concrete is very slow, so tests generally combine diffusion and migration in an electric field. The results presented in section 4.4.1.2 are from this test.

4.4. Degradation mechanisms

4.4.1. *Reinforcement bar corrosion risk*

Reinforcement bar passivation is one of the principal functions of concrete cover. According to Eurocode 2, coating concrete must ensure "good transmission of bonding forces, protection of steel against corrosion (durability) and a suitable fire resistance" (EN 1992-1-1, section 4, 2004). The minimum concrete cover $c_{min, \, dur}$ required for durability depends on the structural class, linked to the lifespan of the project. Specifications for concrete cover depend on exposure classes, defined by Standard EN 206-1 [NFE 04]. These enable the specific nature and the aggressiveness of the environment to which the structures or part-structures are exposed to be taken into account.

When concrete is exposed to atmospheric carbonation and to ingress by chloride ions (which come from de-icing salts or from seawater), the passivation ensured by the cover is altered. Durability is understood as the time required for an attack on the concrete by aggressive agents (carbon dioxide and chloride ions) up until the first set of reinforcement bars is reached, or as the initiation time for corrosion. For SCC as for conventional concrete, this time depends closely on the concrete's physico-chemical properties, and as such the formulation can therefore be adapted to suit the exposure conditions.

4.4.1.1. *Carbonation*

4.4.1.1.1. Degradation mechanism

Carbon dioxide CO_2 which penetrates the concrete dissolves in the interstitial solution and reacts with portlandite $Ca(OH)_2$ and hydrated calcium silicates, forming calcium carbonate $CaCO_3$. The reaction induces a reduction in interstitial solution pH, and the reinforcement bar steel is no longer passivated. Degradation arises essentially from the reaction with portlandite [PAP 91, THI 06] according to the balance equation [4.11]:

$$Ca(OH)_2 + CO_2, H_2O \rightarrow CaCO_3 + 2H_2O \qquad [4.11]$$

Carbon dioxide is naturally present in air and gets into concrete via the porous network and cracks, by a diffusion phenomenon (chemical species transport due to a concentration gradient). The molar volume of carbonation products is larger than that of the reactants, and this gives rise to a reduction in porosity. In addition, water liberated by the reaction leads to an increase in the degree of saturation of concrete. The carbon dioxide migration speed is much lower in water than in air, since the CO_2 diffusion coefficient in the aqueous phase is 10^4 times lower than in the gaseous phase [CHA 97]. Overall, this results in a lower diffusion of carbon dioxide in the concrete, and as portlandite becomes increasingly less accessible as a result of calcium carbonate formation, the speed of the carbonation front reduces as it progresses.

In order to attempt a description or prediction of the carbonation progress in concrete, the diffusion of CO_2 in the concrete must therefore be determined via the CO_2 diffusion coefficient which is a function of the porosity and the relative humidity, and the degree of saturation of concrete.

Accounting for these numerous physical and chemical phenomena leads to a set of models of differential equation complex systems over time and space. However, some simplified hypotheses can be made, especially in that the characteristic time for chemical reactions is low before the CO_2 diffusion in the concrete [PAP 91], even if in reality the carbonation front is not as "straight" as was assumed [THI 06]. Hence, the progression of the carbonation front in concrete is often represented by a square root function time (equation [4.12]):

$$X_c = A\sqrt{t} \qquad\qquad [4.12]$$

where:

– X_c is the thickness of carbonated concrete;

– A is a constant determined from experiments or predictive models.

This progression seems to be well verified experimentally, and the value of the coefficient A can thus be used to compare the resistances of different concrete mixtures to carbonation.

Determining the carbonation resistance of a concrete is carried out by measuring the thickness of the carbonated zone, after one or several time periods. There is a wide choice a priori of methods for estimating carbonated concrete thickness, but the most frequently phenolphthalein is used, a pH indicator whose use is recognized in the operational methods recommended by AFPC-AFREM [AFP 97] and the test standard. Its turning point, around 9, enables carbonated concrete to be distinguished from the unaltered concrete which is colored (Figure 4.9).

In naturally occurring exposure conditions, carbonation is a slow process; the first signs often appear after several tens of years as rust marks on the facing or spalling of the concrete cover (Figure 4.8).

In the design and specification phase of a concrete mixture, ageing tests are sometimes necessary, more or less accelerated and in controlled conditions.

Figure 4.8. *Concrete cover spalling as a result of reinforcement bar corrosion (photo GeM)*

Figure 4.9. *SCC carbonation depths $V_A / V_C = 0.4$ fly ash (on the left) and $V_A / V_C = 0.4$ limestone filler (Table 4.5) after 30 months of natural carbonation 20°C, 50% RH [KHO 07]*

The CO_2 content of ambient air is between 0.03 and 1%. It is in these conditions that natural carbonation occurs. Accelerated carbonation tests use mixtures of gases that are clearly richer in CO_2. The lowest increase in CO_2 for this type of test is 3% [PAP 00]. For the most commonly used test in France, the gas mixture which circulates in the carbonation enclosure contains 50% CO_2 and 50% air.

Accelerating the test is achieved by involving two parameters: the water degree of saturation of concrete and the CO_2 content. CO_2 gets into the concrete more easily when it is gaseous, but the hydrate carbonation reactions can only take place in aqueous solution. Carbonation is therefore negligible for a completely dry or completely saturated concrete, the carbonation speed is optimum for intermediate degree of saturation of concretes – seemingly around 60%, but this can vary according to which source is consulted – from whence the relative humidity rate is recognized in the recommended operational modes (65%).

4.4.1.1.2. Composition and casting parameters which influence concrete carbonation resistance

Even though no model exists which would enable easy correspondence between the composition and performance of a material, numerous experimental results have been published. These allow tendencies to be drawn out and the relevance of certain indicators can be determined.

Composition

Despite the most common approach being prescriptive, numerous studies have tended to investigate material performance criteria. The limit values in the standards come from the fact that they are involved, at the same time, in the level of compactness by the water/binder ratio and in the chemistry with regard to the proportion of Portland cement in the mixture.

Water / cement or water / binder ratio

The water/cement ratio (W/C) influences numerous concrete characteristics, including transport properties, which durability depends upon. The carbonation depth increases with W/C ratio. Also, the carbonation depths measured after 6 years of exposure show a quasi-linear relation [BAR 96], which is probably due to a dilution of the basic content of the binder and an increase in porosity.

Cement and mineral admixture proportions

It appears that the carbonation depth after a given duration is a little sensitive to the cement content, up until a threshold value below which it increases rapidly [BAR 96]. But other studies on the influence of this parameter have not confirmed the existence of a threshold at around the limits of the standard [VEZ 99].

Some studies warn against using high proportions of mineral admixtures, in particular fly ash binder content more than 30% [BAR 96] or more than 50% slag. The negative influence of these mineral admixtures on carbonation resistance has been confirmed by other studies, but it has been shown that its effect can depend strongly on how long it spends in humid curing. Limestone and silica fillers seems to have a positive influence in the case of extended curing (28 days), and a negative influence when the structure is exposed too quickly to drying out, i.e. at around 30% average resistance at 28 days [COQ 99].

Mineral admixtures can actually be involved *a priori* at three levels: diluting the lime in the binder, causing a reduction in the initial amount of portlandite; consuming portlandite CH by pozzolanic reaction (equation [4.13]); and, reducing compactness as a result of a low activity index at early age.

$$S + x\,CH + (y - x)\,H \rightarrow C_x SH_y \qquad\qquad [4.13]$$

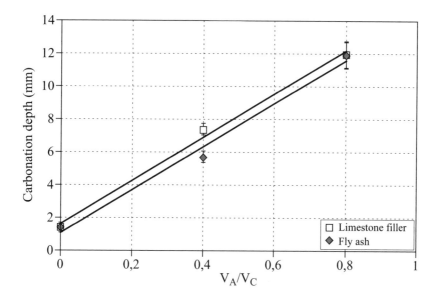

Figure 4.10. *Carbonated depth of SCC after 30 months of natural carbonation at 20°C, 50% RH as a function of the mineral admixture volume/cement volume ratio (V_A/V_C) [KHO 07]*

The carbonation depth has been observed to diminish when the substitution rate of aggregates by mineral admixtures such as silica fumes or fly ash increases, but it increases with the substitution rate of cement [PAP 00]. The carbonation depth also depends on the nature of the admixture, in particular on the fly ash CaO content. The higher the fly ash CaO content, the lower the carbonation depth.

In a study of concrete mixtures with high proportions of ashes [BUR 06], the influences of curing and composition on the carbonation depth have been demonstrated. Furthermore, identical carbonation depths (Figure 4.10) have been shown for concrete mixtures with the same volumetric substitution rates of fly ash and limestone filler (without pozzolanic activity). These results confirm the dilution hypothesis, but not the consumption of portlandite by pozzolanic reactions.

Curing

Several studies demonstrated an interest in extended curing to increase carbonation resistance in concrete [BAR 96, VEZ 99, COQ 99, ASS 04]. Extended curing results in better hydration in concrete cover, and therefore to a closing up of the porosity and a reduction in the CO_2 diffusion coefficient. This sensitivity to the curing mode can be amplified by increasing the proportion of mineral admixtures used to substitute cement (with a constant proportion of water in the mixture).

Compressive strength

Compressive strength is not in itself a durability indicator – two concrete mixtures with the same strength can perform very differently – but it is measured systematically since it forms part of the specifications in the standard. It is closely linked to porosity, and it is therefore interesting to relate it to carbonation. This has been done more or less explicitly in numerous studies.

Generally speaking, the carbonation depth decreases when the compressive strength increases [PAP 91, BAR 96, ASS 04]. For a given compressive strength, the carbonation depth can always vary significantly, as a function of the type of cement used.

Carbonation resistance of a concrete structural element depends on the concrete cover porosity. Under the hypothesis that hydration of the surface concrete cannot continue after the end of the curing, the compressive strength at the end of the curing is a good indicator of its carbonation resistance. The carbonation depth has been shown experimentally to

correlate well with the compressive strength at the end of
the curing (Grandet *et al.* in [BAR 96]).

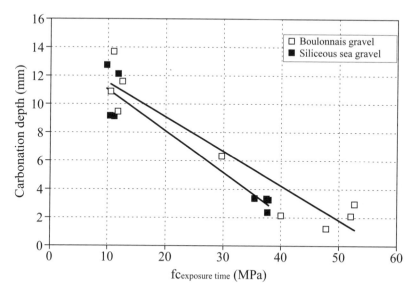

Figure 4.11. *Carbonation depth after one year of carbonation at 20°C, 50%
RH, as a function of the strength at the end of curing [ROZ 09]*

4.4.1.1.3. SCC carbonation

Generally, comparisons between SCC and conventional
concrete reveal similar carbonation kinetics. The observed
differences are linked to the known influence of composition
parameters already described: the water/cement or water/eq.
binder ratio (taking into account the activity of mineral
admixtures), and the proportion of cement in the mixture.

The studies cited do not allow any conclusions to be drawn
on the influence of parameters more specific to SCC, such as
the influence of the paste volume or the super-plasticizer
content, even if an increased paste volume is assumed to
lead to an increase in the CO_2 diffusion coefficient.

If the paste volume increase is obtained by increasing the proportion of cement in the mixture, this results in a reduction of the effective water/cement ratio. On the physico-chemical level, a higher basic content is caused by the clinker in Portland cement (CaO, NaO), and a lower diffusion coefficient.

In this case, a reduction in the carbonation kinetics is observed [SAK 98, AUD 03a, ASS 04]. In the work by Assié, comparisons are made between SCC varieties with constant cement proportions (315, 350 and 450 kg/m^3), and difference effective water/cement ratios. Hence, a constant effective water/cement ratio, the differences between SCC and conventional concrete are not significant. Some SCC mixtures undergo carbonation more quickly, but this is due to a higher effective water/cement ratio, which is also confirmed by the porosity, which is also measured.

4.4.1.2. *Chloride penetration*

4.4.1.2.1. Mechanism for chloride penetration in concrete

Chloride penetration in the natural environment takes place under the effect of two mechanisms: capillary absorption and diffusion. Capillary absorption arises when the dry or partially saturated concrete soaks up saline solution. Diffusion is the result of a chloride concentration gradient in the solution in the pores between the exposed surface and heart of the structure; it arises in the saturated region. In the case of drying–wetting cycles, the two mechanisms can coexist. This involves, for example, inter-tidal zones which are partially immersed (exposure class XS3). The effects of absorption cannot be ignored, since it can lead to concentrations of the same order of magnitude in a shorter time [FRA 01].

When the chloride concentration at the reinforcement bars reaches a critical concentration, often expressed as a ratio of concentration [Cl-]/[OH-], corrosion of the reinforcement bars is likely to begin.

4.4.1.2.2. Measuring the diffusion properties

In making comparisons between materials, it is important to know the absorption and diffusion properties. Whilst the water absorption measurement is the object of a standardized operational mode (see annex G of Standard NF EN 13369), this is not the case for the diffusion coefficient. Since natural diffusion in concrete is generally too slow to characterize compositions, numerous operational modes aim to accelerate it, particularly by setting up an electric field – this is known as migration. According to the mechanisms in play, and the use which will be made of the indicator obtained, the appropriate operational mode must be chosen and understood.

Migration tests

Migration governs ion transport under the influence of an electric field. It is also concerned with a generalization of diffusion transport.

Methods using a permanent or stationary flow rate generally enable a steady state diffusion (or migration) coefficient D_{ssd} (or D_{ssm}) to be determined, and methods with a variable flow rate allow the non-steady state diffusion (or migration) coefficient D_{nssd} (or D_{nssm}) to be determined. Using these values as input data for predictive models is not particularly recommended, but the associated tests are sensitive enough to be able to compare materials.

Migration tests with constant flow rate

A steady state migration coefficient D_{ssm} can be obtained via a migration test under the influence of a difference in electric potential. Numerous operating modes exist, however some of the more commonly cited modes are [TAN 96] and [TRU 00]. Nevertheless, the principle – measuring the flux of chloride ions in the compartment after the cell in order to deduce the diffusion coefficient from it – and the theoretical justification are identical.

Variable flow rate tests

The disadvantages of the steady state diffusion tests lead to the use of a non-steady state rate test to characterize a concrete using a comparable approach. In any case, non-steady state tests are considered to be more representative with regard to real conditions of chloride penetration [GRA 07], since they take into account the bonding of ions to the cement matrix.

A compromise between test duration and representativeness seems to be possible to achieve via the operating mode defined in NT BUILD 492 [NTB 99], also known as the CTH method, or Tang Luping test [TAN 96]. This method is also the object of recommendations in the synthesis of studies in the GranDuBé project (2007). It enables a non-steady state migration coefficient to be determined.

The test involves applying a potential difference across two faces of a concrete sample in order to make chloride ions penetrate the sample by migration. At the end of the test, the sample is cut and a colored indicator based on silver nitrate is used to determine the depth to which chloride ions have penetrated. The non-steady state migration coefficient D_{nssm} is deduced from this value.

4.4.1.2.3. Influence of the concrete composition

For a given binder, the increase in compactness, expressed in terms of the water/cement, water/binder, or W_{eff}/eq. binder ratio leads systematically to a reduction in the steady state diffusion coefficient for chloride ions [ALE 99, GRA 07]. The most influential parameter, however, appears to be the type of binder used. Hence, in the tests compared in the GranDuBé project, it appeared that replacing Portland cement CEM I 52, 5 N with a cement composed of CEM III/C 32, 5 N (with 85% slag), with an equivalent strength (30 MPa) allows a reduction in the apparent and steady state diffusion coefficients at a higher proportion to that which results from an increase in strength (from 30 to 60 MPa), with CEM I cement. Alexander [ALE 99] and Moon *et al.* [MOO 05] have also shown interest in cements and binders with high proportions of slag.

The substitution of Portland cement by fly ash also seems to be effective, in the laboratory [THO 99, TAN 01] and on site. Hence, the effect of high proportions of ash is not significant at 28 days but appears after several months (Table 4.3).

The characterization of compositions at 28 days does not take such behaviors into account. Hence the Canadian standard CSA A23.1-04 specifies limit values at 56 days, to take binders with slag or ash into account. Kinetics of the evolution of the non-steady state diffusion coefficient have in any case been proposed by several authors:

$$D_t = D_{28} \left(\frac{t_{28}}{t} \right)^m \quad \text{[THO 99]} \quad \text{or} \quad D(t) = at^{-b} \quad \text{[TAN 01]}$$

[4.14]

The positive effect of silica fumes has been shown by the same authors. However, silica fumes seem to have an effect primarily before 28 days, and not on the longer-term kinetics (Table 4.3). Otherwise, this type of chemical additive would limit the sensitivity of early exposure to drying, which is sometimes hampered for binders which contain ash or slag, for which the activity coefficient stays low during the first few days [ALE 99].

	D_{28} (10^{-12} m²/s)	m
Control	4.3	0.23
Fly ash 25%	4.4	0.62
Cement type 10SF (*)	0.66	0.25
Cement type 10SF+fly ash 25%	0.37	0.40
(*) Type 10SF: Portland cement ground with 8% silica fumes		

Table 4.3. *Non-steady state diffusion coefficients for concrete with silica fumes and fly ash [THO 99]*

From this data, the interest in ternary binders ash or slag + silica fumes is threefold: reduced sensitivity to curing mode, characterization of composition at 28 days, long-term effectiveness.

Finally, to the best of our knowledge no experimental study on varying the paste volume (water + binder) with constant proportions exists, but insofar as chloride penetration occurs in the paste, this parameter is likely to be influential. In any case, such variations may explain how the overall porosity does not link well to the diffusion coefficient, and furthermore the fact that it takes neither the network connectivity nor complexity into account.

4.4.1.2.4. Comparing SCC and conventional concrete

Numerous results indicate equivalent behavior in SCC and conventional concrete from the point of view of diffusion [MOR 01, ATT 02, AUD 03b, TRA 03]. This confirms the secondary influence of parameters such as the paste volume and super-plasticizer content.

In the absence of systematic studies, it is, however, difficult to decouple the influences of different parameters. Hence, in order to obtain different SCC compositions, Assié [ASS 04] varied the effective water content (W_{eff}), and hence the paste volume, while keeping the cement proportions constant. The variations obtained were not significant, except in the case of SCC40 which performed better than the corresponding conventional concrete. This did not arise from the influence of the effective water/cement ratio, equal or higher for SCC than for conventional concrete. It must be noted that in this study the SCC compositions were different from the conventional concrete compositions on account of their significant filler contents (150 kg/m^3 for SCC20 and 140 kg/m^3 for SCC40).

This tends to reduce or reverse the composition differences in terms of the effective water/eq. binder ratio (with regard to differences in W/C), since the eq. binder allows some of the limestone filler to be taken into account with a coefficient of 0.25 (from contribution criteria to the compressive strength). The differences in diffusion are not significant (for SCC20 and SCC60) or in favor of SCC (SCC40), we can therefore assume that limestone filler is involved at the micro-structural level, which can be observed in measurements of chloride ion diffusion.

The main parameter seems to be the type of mineral admixture used [ZHU 01, ZHU 03]. For reasons of controlling the rheology, limestone filler seems to be the most widely used. But the incorporation of fly ash enables, as is the case of conventional concrete, a significant reduction in the chloride diffusion coefficient.

4.4.2. Aggressive water

Leaching is a concern for all immersed concrete structures. The interstitial solution in the concrete is actually highly basic and charged with calcium and alkaline ions, different to most water in contact with the concrete, which sets up a steep concentration gradient. Calcium diffusion leads to dissolution of the hydration products in the cement matrix which can be coupled with penetration of species which can induce other types of material and structural degradation, such as sulfate, magnesium, chloride and carbonate ions [GUI 04].

Leaching occurs more intensely when the water in contact with the concrete is a little mineralized and acidic. Degradation propagates from the surface and results in a lowered pH, increased porosity, degraded mechanical properties, and increased transport properties likely to amplify possible degradations, which are coupled. Leaching and acid attacks are first of all of interest, then external sulfate attacks, and the degradation mechanisms are linked [ROZ 09].

4.4.2.1. *Leaching and acid attacks*

Due to its causes and consequences, leaching is the subject of a large number of studies of concrete products: networks, storage and treatment of drinking water, waste water and rain water, hydroelectric structures, radioactive waste storage, and concrete in agricultural exposures.

Since these environments are considered to be harsh for concrete, and since stakes are often high, the prescribed ways of designing concrete mixtures are relatively constrained. Nevertheless, numerous experimental studies provide data which enables the phenomenon to be understood and tools which allow evaluation of material performance other than by their composition, and, to a certain extent, predictions of their ageing.

4.4.2.1.1. Degradation mechanism

Leaching of paste, mortar and concrete

Aggressive environments, from the point of view of leaching, can be quite varied, and combine other types of attack. But water in contact with concrete is slightly mineralized, in particular with calcium, and is neutral (pH=7) to acidic (pH<7). The degradation mechanism induced comes from these concentration gradients between the attacking water and the interstitial solution in the concrete.

Since the interstitial solution is no longer saturated with calcium and hydroxide ions, the solid components containing these species dissolve. The calcium content of solid products decreases with reductions in the calcium concentration in the solution in the pores. Leaching begins through dissolving portlandite, then mono-sulfoaluminates, ettringite and C-S-H decalcify. This process has been demonstrated experimentally for cement paste (Figure 4.12) and concrete.

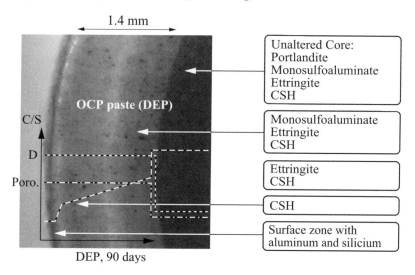

Figure 4.12. *Aging test in pure water on cement paste [GAL 05]*

Dissolving portlandite and decalcifying hydrated calcium silicates (C-S-H) leads to an increase in porosity and in the diffusion coefficient [MAI 00]. The progression of the degradation results in advancing dissolving fronts which can be observed using colored pH indicators or by microscopy (Figure 4.12).

The experimentally observable quantities are the outward flux of calcium and hydroxide ions, and the advancing of the portlandite dissolving and hydrate transformation front. So long as the pH of the solution in the pores is essentially governed by the presence of portlandite $Ca(OH)_2$, the advancement of the dissolving front results in a fall in pH, which can be observed by indicators such as phenolphthalein.

As soon as the alkalis diffuse towards the outside of the paste, the kinetics at the forefront of total portlandite dissolution, the same as the quantities of ions leached, follow an evolution close to a square root function of time, which is described experimentally and modeled.

This description of the degradation mechanism for cement pastes is applicable to concrete, but the aggregates are involved at several levels and make predicting the degradation kinetics more difficult:

– influence of the paste volume;

– development of interface transition zones;

– aggregate geometry;

– mineralogical type (leaching can occur from aggregates);

– cracking due to paste shrinkage being prevented by aggregates.

Ageing tests

Two main preoccupations direct the development of ageing tests: representativeness and acceleration of degradation. The first objective can be achieved by controlling the boundary conditions, and by comparing the accelerated degradation facies with those from a natural or non-accelerated degradation which corresponds to the environment studied [BAD 06].

The test acceleration is obtained using a solution without calcium and by increasing the pH gradient and/or the concentration of aggressive agents – acids (for example ammonium ions NH_4^+). These preoccupations have led to the development of two types of tests: leaching in the presence of ammonium nitrate and leaching in pure water (degradation in pure water) in acidic solution.

In the test with so-called "pure" water, acidic or otherwise, acceleration of the test comes about only from the calcium concentration gradient and pH – i.e. hydroxide ions [BOU 94, PLA 06]. The solution is refreshed and pH is controlled – by the addition of nitric acid in the Planel pure water degradation (nitrate ions are not involved in degradation). The amount of acid added hence gives the quantity of hydroxide ions leached, and the solution concentration gives the quantities of calcium ions leached. This enables the degradation kinetics to be determined, and situated in representative conditions. The ammonium nitrate leaching test consists of immersing the concrete in a 1 mol/L solution of ammonium nitrate. Leaching hydroxide ions results in an increased pH. When it reaches 8.3, the solution is refreshed. It is also possible to dose the solutions and to obtain leaching kinetics for calcium ions. The corresponding balance equation for the reaction between ammonium nitrate and portlandite is as follows:

$$2\, NH_4NO_3 + Ca(OH)_2 \longrightarrow Ca(NO_3)_2 + 2\, NH_3 + 2\, H_2O$$
$$[4.15]$$

Insofar as ammonium ions are involved in the acid-base degradation reactions, the difficulty for the operating mode is in avoiding the formation of expanding phases inside the sample, such as calcium nitroaluminates $(3CaO.Al_2O_3.Ca(NO_3)_2.10H_2O)$ which can induce splitting in the concrete.

Nevertheless, these two types of tests are based on the same degradation mechanism [CAR 96, MOR 04]. The leaching test in the presence of ammonium nitrate can lead, in certain conditions, to leaching kinetics ten times as fast as leaching in pure water – non-acidified – since that appears in the reference (where the kinetics are 1.56 and 0.140 mm/j$^{1/2}$ respectively). The kinetics depends on the ammonium nitrate concentration: in the literature the concentration is between 0.5 and 48% ammonium nitrate per liter of solution.

Properties affected by leaching

Leaching affects concrete durability both of the material and of structures. The most direct consequences are the increase in porosity and transport properties, and the lowering of interstitial solution pH, which can lead to initiating corrosion of the steel reinforcement bars.

The loss of mechanical strength induced by the increase in porosity of leached concrete has been demonstrated in many experimental studies [CAR 96, LEB 01, NGU 06] from leaching in ammonium nitrate. During some of these tests, the coupling between the effects of mechanical and chemical disturbances has been observed. Finally, leaching can be coupled to sulfate penetration, which causes expanding reactions [GAL 05, PLA 06].

4.4.2.1.2. Influence of concrete composition

Influence of compactness

Since the dominant transport mechanism is diffusion, one of the approaches adopted for reducing the risk of diffusion is to reduce the diffusion coefficient by increasing the compactness – often formulated in terms of the water/cement ratio or the water/binder ratio. For a given binder, experimental studies have revealed a quasi-linear reduction in leaching kinetics with this ratio [KAM 03, MOR 04, BAD 06].

The results of Assié's study [ASS 04] on self-compacting and conventional concrete mixtures with analogous strengths (B20, B40, B60) come from the leaching test in ammonium nitrate. The kinetics are comparable for the same strength class and increase with reductions in compactness, which can be expressed in terms of the W_{eff}/eq. binder ratio, the strength, and the porosity or effective chloride diffusion coefficient. However, the conclusions of the studies with a constant binder cannot be extended to materials with different binders. In effect, a concrete based on CEM V with higher W/C ratio and porosity than a concrete based on CEM I can have a better leaching resistance (Table 4.4), according to the pure water degradation test [BAD 06].

	Concrete 1	Concrete 2	Concrete 3	Concrete 4
Cement (kg/m³)	409	408	405	356
Cement type	CEM V/A 32. 5 N	CEM I 52, 5 N	CEM II/A-D 52, 5 R	CEM I 52, 5 N
W_{eff}/C	0. 39	0. 33	0, 24	0,.31
fc $_{90d}$(MPa)	61	60	91	72
Porosity at 90 d (%)	13. 7	11. 3	7, 4	8. 5
Leaching test at constant pH				
Total amount of calcium ions leached at 60 d (mmol/dm²)	10. 9	39. 2	13. 5	24. 7
Equivalent degraded thickness [EPD_{eq}] (mm)	0. 4	0. 8	0. 3	0. 6
Leaching test with ammonium nitrate				
Total amount of calcium ions leached at 60 d. (mmol/dm²)	165	318	178	257
Equivalent degraded thickness [EPD_{eq}] (mm)	5. 4	6. 8	4. 1	6. 3

Table 4.4. *Amounts of calcium ions leached and equivalent degradation thicknesses associated with leaching tests at constant pH or with ammonium nitrate [BAD 06]*

Influence of the binder

As indicated before, the type of binder used can have a significant influence on concrete performance.

Binders that are low in CaO are often recommended since they are involved at two levels in the leaching mechanism: on the one hand, they produce a lower portlandite content – the dissolution of which leads to increased porosity – and denser, more stable C-S-H; on the other hand, with an identical overall porosity, the morphology of the pore network is different, and this results in lower capillary porosity and diffusion coefficients [KAM 03]. Binders which include slag or pozzolanic additives, such as silica fumes and fly ash can give rise to a better resistance to leaching [CAR 96, BAD 06].

4.4.2.1.3. Studies on SCCs

Little data is available in the literature. We rely primarily on results from Assié [ASS 04] and Rozière [ROZ 09, ROZ 11].

In Assié's research, where comparisons are made while the cement proportion is constant, the concrete specimens are immersed in an ammonium nitrate solution. The behaviors of SCC and conventional concrete are, generally speaking, equivalent, taking into account the scatter in the results. An interesting example of a significant difference is concerned with SCC40 and CC40 (Figure 4.13).

The leaching kinetics are classified in the following way: CC 40 III < SCC40 III < CC 40 I. The respective W_{eff}/C ratios are as follows: 0.50 – 0.55 – 0.60. This seems to explain the observed classification. Yet leaching kinetics are closely linked to diffusion, and limestone filler seems to have a positive effect at this level (see the chloride ion diffusion coefficient) and the W_{eff}/eq. binder ratio is therefore more relevant when comparing compositions.

The SCC40 and CC40 concrete mixtures have the same W$_{eff}$/eq. binder ratio of 0.50; their leaching kinetics are identical until 56 days, then the kinetics of SCC40 increase. The limestone filler can be assumed to be affected by leaching, since it is essentially made of calcium carbonate, which is soluble in acid environments. This is confirmed by other results.

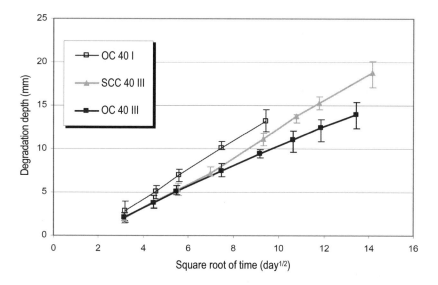

Figure 4.13. *Leaching in ammonium nitrate [ASS 04]*

The test to which the following concrete mixtures are subjected [ROZ 07] (Table 4.5) is the leaching test at constant pH. The cumulative amounts of hydroxide and calcium ions leached allow the leaching kinetics to be calculated. After immersion for 60 days, a 1% phenolphthalein solution is ground onto a fresh longitudinal crack in the sample, to determine the zone which has been degraded from the extent of the coloring.

(kg/m^3)	A = Limestone filler			A = Fly ash	
V_A/V_C	0	0.80	0.40	0.80	0.40
Limestone gravels					
3/8 mm	790	790	790	790	790
Sand					
0/4 mm	670	670	670	670	670
Cement (C)					
CEM I 52.5 N	663	365	470	365	470
Mineral admixture (A)	0	255	165	203	131
W_{eff}	204	199	198	199	200
Super-plasticizer	12.22	5.20	4.84	6.66	8.59
Thickening agent	0.66	0.66	0.66	0.66	0.66
Binder (C+A)	663	620	635	568	601
W_{eff} / binder	0.31	0.32	0.31	0.35	0.33
Eq. binder (C+k.A)	663	365	470	459	549
W_{eff} /eq. binder	0.31	0.54	0.42	0.43	0.36
A/(A+C)	0	0.41	0.26	0.36	0.22
Paste volume Vp (L/m^3)	420	415	415	415	417
Slump flow (cm)	61	80	76	78	68
fc$_{28}$ (MPa)	64.1	44.8	52.3	54.5	60.1

Table 4.5. *Self-compacting concrete compositions*

The ratios of leaching kinetics for hydroxide and calcium ions are close to 2, which is *a priori* in agreement with the chemical reaction coefficients for the dissolution of portlandite in acidic conditions (equation [4.16]). The ratios are, in fact, higher than 2, and some of the calcium ions may come from the CSH. However, this approach does not enable the dissolution of calcite CaCO₃ (from aggregates and limestone fillers, or produced by carbonation) to be excluded, which also requires two moles of H₃O⁺ ions per mole of calcite (equation [4.17]).

$$Ca(OH)_2 + 2H_3O^+ \rightarrow Ca^{2+} + 4H_2O \qquad [4.16]$$

$$CaCO_3 + 2H_3O^+ \rightarrow Ca^{2+} + H_2O, CO_2 + 2H_2O \qquad [4.17]$$

Comparisons between the kinetics cannot therefore be used except between concrete mixtures made with the same types of aggregates.

Leaching kinetics

With the same admixture content, the dissolution kinetics for SCC with limestone fillers are noticeably faster than for those with ashes. This may be due to a higher diffusion coefficient or dissolution of limestone filler.

(kg/m³)		A = Limestone filler		A = Fly ash	
V_A/V_C	0	0.80	0.40	0.80	0.40
Hydroxide ions OH⁻ (mmol/dm²/d^{1/2})	9.24	10.46	10.41	8.33	7.75
Calcium ions Ca²⁺ (mmol/dm²/d^{1/2})	5.08	5.91	-	5.06	-

Table 4.6. *Leaching kinetics*

Degraded thicknesses

At the end of the test, the sample is pulled out of the solution and the degraded thickness is measured using a phenolphthalein solution and the same principle as for measuring the depth of carbonated concrete. The results (EPpH) are given in Table 4.7. The measured values are low. In order to obtain a more precise indicator, the degraded thickness is also estimated (EPDeq) from the total quantity of leached ions during the test and the initial calcium content of the binder – or of the cement [BAD 06] – according to equation [4.18]:

$$EPDeq = \frac{Total\ leached\ calcium}{Initial\ calcium\ content\ in\ binder} \qquad [4.18]$$

Microstructural analysis of the samples subjected to the same type of test [BAD 06] tend to show that the equivalent degraded thickness (EDPeq) underestimates the leached thickness, since the degraded zone is not completely decalcified, which is an assumption in the previous calculation.

The data collected in Table 4.7 enable comparison of EPpH and EPDeq values. The EPDeq overestimates the degraded thickness, because calcite $CaCO_3$ also dissolves from aggregates and limestone filler. Photos of three samples of self-compacting concrete (Figure 4.14) show the degradation front, revealed by a phenolphthalein solution. These photos show that in all three cases the degraded thickness was underestimated by the EPDeq calculation.

In photos 13.a and 13.c in Figure 4.14, a drop away from the interface can actually be seen at the aggregate level.

(kg/m³)	A = Limestone filler			A = Fly ash	
V_A/V_C	**0**	**0. 80**	**0. 40**	**0. 80**	**0. 40**
Total quantity of calcium ions leached at 60 days (mmol/dm²)	32. 5	35. 7	35. 0	33. 3	30. 0
Initial calcium content (mol/dm³)	7. 70	4. 24	5. 46	4. 31	5. 50
Estimated amount of leached calcium ions (aggregates) (mmol/dm²)	23. 7	23. 7	23. 7	23. 7	23. 7
Equivalent degraded thickness EPDeq. (mm)	0. 42	0. 84	0. 64	0. 77	0. 55
Equivalent degraded thickness EPDeq.2 (mm)	0. 11	0. 28	0. 21	0. 22	0. 11
Degraded thickness (phenolphthalein) EPpH (mm)	0	0. 5	0	< 0. 5	0

Table 4.7. *Degraded thicknesses*

In order to explain the disagreements between the EPpH and EPDeq values for concrete made with limestone aggregates, the EPDeq2 calculation (equation [4.29]) takes into account the drop away from the interface at the level of limestone aggregates.

$$EPDeq2 = \frac{Total\ leached\ calcium - Leached\ calcium\ from\ aggregates}{Initial\ calcium\ content\ in\ binder}$$

[4.19]

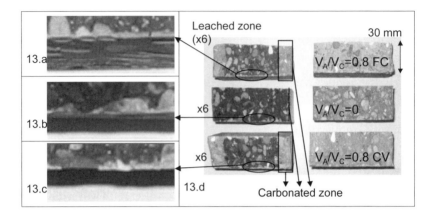

Figure 4.14. *Revealing degraded zones using a pH indicator*

The amount of leached calcium which comes from the aggregates (in mmol/dm²) was calculated by taking into account the volumetric proportion of limestone aggregates – Vol. (G. 3/8 mm) for self-compacting concrete – which is assumed to be made of 100% calcite, and by varying the leached thickness (drop away from the interface) in equation [4.20]:

$$Leached\ calcium\ (aggregates) = \frac{drop(G)\ P_G\ \rho_G}{M_{CaCO_3}}$$

[4.20]

– P_G is the volumetric proportion of limestone aggregates;

– ρ_G is the volumetric mass of aggregates;

– M_{CaCO_3} is the molar mass of calcium carbonate.

The values for EPDeq2 given in Table 4.7 were obtained using a drop away in the interface of 0.3 mm for aggregates in SCC, which is in agreement with the characteristics of the degraded samples. These EPDeq2 values are not predictive values but agree with the EPpH values, except for the concrete with $V_A/V_C=0.8$ limestone filler.

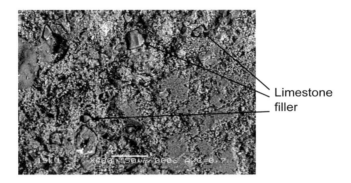

Figure 4.15. *SEM photo of the leached zone of the sample* $V_A/V_C=0.8\ FC$

The concrete with $V_A/V_C=0.8$ limestone filer degraded more rapidly than the concrete with $V_A/V_C=0.8$ fly ash, even though the constituent volumetric proportions are the same. Only the type of mineral admixtures changed: limestone filler for the first and fly ash for the second. Observation via a scanning electron microscope confirms the negative effect of limestone filler on the behavior of concrete (Figure 4.15). In this photo of the degraded zone, a dissolved ring can be seen around the filler grains.

Conclusion

Even though they do not completely prevent performance evaluation, limestone aggregates appear to be little suited to exposure to acidic environments. This has been shown in real exposure situations (hydroelectric works) where leaching of such aggregates can be quicker than that of the paste. It therefore seems reasonable not to use them in severe exposure conditions. The use of limestone filler, as studied, appears not to be very desirable since it is likely to speed up degradation. On the other hand, substituting cement with other mineral admixtures in amounts to keep the binder constant leads to an equal, if not better, performance. SCCs with fly ash have comparable performance characteristics to conventional concrete mixtures with comparable compactness (W_{eff}/C ratio) and binder (cement + fly ash) E_{eff}/C) [ROZ 09].

4.4.2.2. *External sulfate attacks*

Sulfate attacks are often spoken about in the plural. The use of the plural to describe concrete degradation by sulfates is not trivial, since there is a real multiplicity of attacks, in causes as well as in manifestations of this phenomenon. We focus on the study of sulfate attacks from external sources, i.e. when sulfates come from the environment of the concrete.

Sulfates can come from very varied origins: water and soils rich in gypsum, sulfates emitted by industrial products, from fertilizers or organic matter, from sea water, etc. Standard EN 206-1 on concrete composition considers such environments to be chemically aggressive – exposure class XA in Standard EN 206-1 – and precautions must be taken with regard to the cement choice and concrete mix design. Sea water itself contains sulfates, in the order of a few grams per liter, and concrete in the marine environment is also affected by external sulfate attacks (exposure class XS). The consequences of sulfate penetration in concrete are the formation of gypsum, ettringite, or thaumasite, and the transformation of hydration products. This can lead to swelling, loss in strength and changes in the surface of a concrete structure. If the first of these three phenomena is the best known and most characteristic, the other two are not to be neglected in practice.

If sulfate ions SO_4^{2-} are always the cause [NEV 04], the conditions in which they act, particularly the associated cation, temperature, pH, and concentration strongly condition the degradation mechanism and type. In any case, for the same process and degradation stage, the attack rate varies as a function of the measured indicator [SAH 07].

Few studies exist on the behavior of SCC exposed to external sulfate attacks, but some data are available on the problems common to conventional concrete and SCC: the effects of compactness, cement type and mineral admixtures.

4.4.2.2.1. Mechanism for degradation by sulfates

The modes of action of sulfates on concrete are complex. Nevertheless, mineral chemistry and some recent investigation techniques have allowed recognition of some compounds, explanations of how they form, and suggest ions of scenarios which agree with the observed degradation modes.

Chemical transformations of concrete

As well as the concrete components, and hydration products (C-S-H and portlandite), three compounds, which form in the presence of sulfates, play an important role in external sulfate attacks: gypsum, ettringite, and in certain cases, thaumasite.

For solutions with high sulfate contents – higher than 1,000 mg [BAR 96] or 8,000 ppm [SAN 01] SO_4^{2-} – gypsum is produced from the interaction between the sulfate deposits and the cement matrix. Gypsum, $CaSO_4.2H_2O$, is used as a setting regulator and is present in cement. Therefore, it already exists in sound concrete. The gypsum which forms in the presence of a sulfate-rich solution results from the substitution of hydroxide ions with sulfate ions in portlandite $Ca(OH)_2$, according to reaction [4.21], in the case where the associated cation is sodium Na^+.

$$Ca(OH)_2 + Na_2SO_4 + 2H_2O \rightarrow CaSO_4.2H_2O + 2NaOH$$

[4.21]

Gypsum produced in this way can then react with calcium aluminates to form ettringite, $C_3A.3CaSO_4.H_{30-32}$ (where C indicates CaO, A is Al_2O_3 and H is H_2O), from anhydrous C_3A according to reaction [4.22]. But ettringite can also form from hydrated tetracalcium aluminate C_4AH_{13} or from calcium monosulfaluminate $C_3A.CaSO_4.H_{12}$.

$$C_3A + 3CaSO_4.H_2O + 24 - 26H_2O \rightarrow C_3A.3CaSO_4.30 - 32H_2O$$
[4.22]

The mode of action of sulfates depends closely on the associated cation. Hence magnesium sulfates have a particular mode of action – even more detrimental to concrete since, as well as forming ettringite, according to reactions [4.23] and [4.24], it also causes the substitution of calcium ion Ca^{2+} with magnesium ions Mg^{2+} in the hydrated calcium silicates C-S-H, according to reaction [4.25]. This leads to a loss of binding properties and therefore to a loss of mechanical strength in the concrete since hydrated magnesium silicates M-S-H are not binders. Nevertheless, magnesium hydroxide $Mg(OH)_2$ in the form of brucite, produced in reaction [4.23], can slow down sulfate penetration and form a protective layer, if it is not mechanically destroyed. This can explain the relatively good behavior of concrete in seawater, however rich in sulfate and magnesium ions that water may be [NEV 04].

$$Ca(OH)_2 + MgSO_4 + 2H_2O \rightarrow CaSO_4.2H_2O + Mg(OH)_2$$
[4.23]

$$C_3A + 3CaSO_4.H_2O + 26H_2O \rightarrow C_3A.3CaSO_4.32H_2O$$
[4.24]

$$C-S-H + MgSO_4 + 2H_2O \rightarrow CaSO_4.2H_2O + (C,M)-S-H$$
[4.25]

Ettringite

The formation of ettringite $C_3A.3CaSO_4.H_{30\text{-}32}$ is often linked to descriptions of sulfate attacks. It forms from C_3A, the lesser phase of cement. Even though it often plays a major role in these phenomena, its presence is not necessary or sufficient to explain swelling and degradation of concrete subjected to sulfate attacks.

Different types of ettringite, or hydrated calcium trisulfoaluminate, must be distinguished [DER 94]. *Primary ettringite* is the product of the reaction between calcium sulfates introduced into the cement as setting regulators and tricalcium aluminate C_3A. This decomposes to form calcium monosulfoaluminate. It is neither expansive nor pathological and is most often presented in the form of needles.

Secondary ettringite develops when the concrete has already hardened. If the sulfates are from external sources, they can give rise to expansion. This is what is caused in the case of external sulfate attacks.

Differed ettringite is formed from internal sulfate attacks, and may or may not give rise to expansion. It is an issue for concrete in which primary ettringite has not been able to form during the first few moments of hydration. Its formation leads to internal strains which can cause swelling and cracking of the concrete. This pathology requires several conditions to coincide, as much to do with the concrete constituents as with the humidity and temperature of the concrete and its environment.

The expansion of ettringite, which is linked to its crystallization mode, can depend on the solution composition, and in particular on its lime CaO content. The solubility of ettringite noticeably reduces with the CaO content [BAR 96]. The products formed have a molar volume three to eight times higher than that of the reactants (C_3A or $C_4A.H_{13}$), which leads to the development of high stresses.

Finally, ettringite formation is often linked to the formation of gypsum, which can also have deleterious and expansive effects.

Gypsum

Gypsum can be produced via reaction [4.21], but it can also result from the dissolution of ettringite, in relatively weak solutions of calcium hydroxide, when the pH is below 11.5-12 [SAN 01].

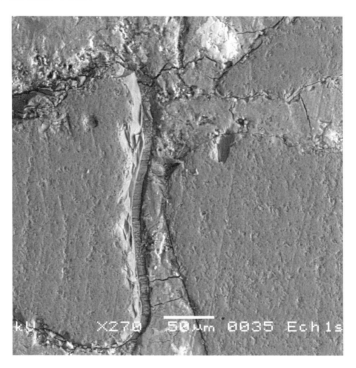

Figure 4.16. *Degradation in a sulfate solution with gypsum formation in the interfaces between the paste and aggregates (SEM observation, GeM)*

Its role in sulfate attacks has not yet been completely determined, but a certain number of results [SAN 01] show that it can lead to degradation by itself (Figure 4.16). The absence of C_3A or ettringite for example, does not preclude a sulfate attack from occurring. The damage caused can be of two types: scaling or swelling of the concrete. To evaluate the consequences of gypsum formation alone, the formation of ettringite must be prevented, by using binders without C_3A.

A study by Mehta *et al.* [MEH 79] shows that gypsum formation can be responsible for scaling and significant mass loss, but the experiments did not include length monitoring. Tian and Cohen [TIA 00] studied the proportional variations of mortar blocks based on C_3S. They caused substantial swellings and demonstrated gypsum formation, without linking the two phenomena explicitly. The amount of gypsum formed can be reduced by substituting C_3S for silica fumes, which confirms its sources as the consumption of portlandite (equation [4.21]).

Thaumasite

Thaumasite, $CaCO_3.CaSO_4.CaSiO_3.15H_2O$, forms preferentially when sulfate attacks occur at low temperatures (between 0 and 5°C). It is the product of reactions between C-S-H, sulfate ions SO_4^{2-} and carbonates CO_3^{2-}. It can also form from ettringite and be associated with the formation of gypsum [SAN 01].

Concrete degradation which is linked to thaumasite formation comes from the degradation of C-S-H. The probability of seeing this degradation type increases when limestone aggregates or fillers are used in cold climates, since these materials are essentially made of calcium carbonate, and carbonate ions are involved in the reactions which form thaumasite.

However, thaumasite formation has also been observed in climates without such low temperatures. In any case, thaumasite is not easy to identify, since it is often linked with ettringite and, in a similar way, its formation is not always detrimental to concrete durability [NEV 04]. Its formation mechanism is not yet well understood, and neither is the influence of temperature on sulfate attacks.

Degradation process

From laboratory studies and detailed observed techniques, scenarios which involve the mechanisms and products previously described have been proposed for the development of external sulfate attacks in concrete.

Irassar *et al.* [IRA 03] put forward a scenario developed from composition profiles determined by X-ray diffraction analysis and scanning electron microscopy. Tests were conducted on mortar samples (25x25x285 mm) using the ASTM C 1012 test type [AST 00]. Concentration profiles for gypsum, ettringite, portlandite and $CaCO_3$ were measured after 1 and 2 years of immersion in a solution of Na_2SO_4 (0.352 mol/L). This showed the attack progression, for Portland cements and binders based on cement and on limestone filler. The proposed system includes the following stages:

– diffusion of sulfate ion SO_4^{2-} and $Ca(OH)_2$ dissolves;

– ettringite formation;

– gypsum formation and reduction in $Ca(OH)_2$ concentration;

– decalcification of C-S-H;

– thaumasite formation.

In this mechanism, ettringite appears as the first sulfate attack product: its formation is explained by the reaction between sulfate ions SO_4^{2-} and monosulfoaluminates or monocarboaluminates. Ettringite is not stable at pH values less than 10.7 [IRA 03] or in the range of 11.5-12 [SAN 01]. In work by Irassar *et al.* [IRA 03], the solution pH was maintained at 7 ± 1. At the surface, ettringite could therefore decompose to from gypsum.

In addition, when the aluminate source was exhausted, ettringite formation ceases and sulfate ions SO_4^{2-} react with calcium ions Ca^{2+} to form gypsum. These conclusions are in agreement with the steps in concentration profiles for ettringite, and with another study conducted with sodium sulfate solution held at pH values < 10.5 and constant composition with daily renewal of the solution.

X-ray spectroscopy and scanning electron microscopy have also shown that thaumasite can form at ambient temperatures (20°C), even though its occurrence is often assumed to be more likely and more significant at lower temperatures. In this study, all the necessary pieces were in place: calcium sulfate, calcium carbonate (in the limestone filler), silicates and water. Its formation also seems to be favored by cements that are rich in C_3A, since they give rise first of all to ettringite, which always precedes thaumasite formation. Thaumasite formation thus forms the final stage of the proposed mechanism and is characterized by different degradations to those induced by ettringite.

Although it can cause swelling and even cracks in the concrete, it does not alter its strength. On the other hand, thaumasite formation affects C-S-H and therefore material cohesion, which can be visually observed.

4.4.2.2.2. Influence of concrete composition

Several laboratory tests concerned with external sulfate attacks, as well as some field data, demonstrate the influence of some parameters on concrete resistance to sulfates. Even though some results seem to contradict one another, a few tendencies and composition ranges can be identified where the risks are higher.

French standardization context

European Standard EN 206-1 [NFE 04] classes sulfate rich regions as chemically aggressive environments – exposure classes XA1, XA2 and XA3. The standard defined three classes as a function of sulfate and magnesium ion concentration in water, and sulfate ions in soil.

In the framework of the prescriptive approach for specification of concrete, Appendix NA.F defines composition limits that are more restrictive than for other chemical attacks with regard to mineral admixtures (A).

The upper limit for the A/(A+C) ratio is *set to 0.15 when the attack comes from the presence of sulfates*, instead of 0.30, for fly ash and ground slag. In any case, limestone or siliceous fillers cannot be used.

As well as this, there is a recommendation *for class XA1, use PM cement* ["seawater resisting"] *and the binder composition must be such that it complies with the restrictions in NF P 15-317 [NFP 95a] and for XA2 and XA3 classes, uses an ES cement ["sulfate resisting"] and the binder composition must be such that it complies with the restrictions in XP P 15-319* [NFP 95b].

These two standards limit the amount of SO_3 and other cement components. Standard NF P 15-317 in particular defines limiting values for C_3A (tricalcium aluminate) and C_3S (tricalcium silicate). Standard XP P 15-319 governs the quantities of C_3A and C_4AF (tetracalcium aluminoferrite):

– $(C_3A) \leq 5\%$

– $(C_4AF) + 2(C_3A) \leq 20\%$

Calcium aluminate C₃A and calcium silicate C₃S contents in cement

Several studies have attempted to demonstrate the effects of the C₃A and C₃S contents of cement on the concrete sulfate resistance of concrete. Even though sulfate resistance increases when the C₃A contents reduces, the results are sometimes difficult to analyze [NEV 04]. Some sulfate degradation mechanisms do not require the presence of C₃A as a necessary condition for activation. This is the case, for example, for thaumasite formation degradation, which occurs preferentially at low temperature in the presence of carbonates: so-called sulfate resistant cements, i.e. those that are low in C₃A, seem to only be a little effective [SAN 01, NEV 04].

It is useful to distinguish different types of sulfate attacks, since some parameters, such as the type of cation, can alter the effect of C₃A content. Therefore, the use of low C₃A cements can have adverse effects on the expected results, in magnesium sulfate or sulfuric acid solutions [NEV 04].

However, for sodium sulfate solution, the degradation kinetics seem to increase linearly with C₃A content, for constant cement content [DUV 92]. However the latter parameter seems to have a stronger effect, and low C₃A content in the cement is not sufficent for the concrete to withstand sulfate attacks well.

The C₃A/SO₃ ratio seems to be a good indicator of resistance to seawater (which contains sulfates) [DUV 92]. In effect, Standard NF P 15-317 [NFP 95a] takes into account the C₃A content to set the maximum SO₃ content, and a ratio less than or equal to 3 for cement guarantees good resistance to seawater resistance.

Several studies have focused on the C_3A content and W/C ratio together [NEV 04], for example the study by the US Bureau of Reclamation, which lasted more than 40 years. They showed that the effect of the C_3A content dominates at high W/C ratios, higher than 0.40-0.45. Beyond these values, concrete permeability increases noticeably with the W/C ratio. It could be thought that when the capillary pores are connected and aggressive agents can penetrate inside the concrete, the chemical resistance of the binder makes the difference.

The C_3S content, and particularly the C_3S/C_2S ratio, also seems to have a noticeable effect on the sulfate resistance, even though few data are available [SAN 01]. C_3S hydration produces more calcium hydroxide $Ca(OH)_2$ than C_2S hydration, because of gypsum formation and C-S-H decomposition which may be induced. This has been observed especially in the case of magnesium sulfate attacks.

Water/cement ratio and cement mixture proportions

Concrete resistance to external sulfate attack increases when the W/C ratio decreases and the proportion of cement in the mixture increases [DUV 92, NEV 04]. In effect, when the W/C ratio reduces, the volume and connectivity of the pore network reduces, and therefore renders it less sensitive to attack since penetration of the concrete is more difficult for the aggressive agents to achieve.

Aggregates

Aggregates are involved at several levels in concrete sulfate resistance. Aggregate skeleton compactness and interface transition zone quality (linked to the aggregate type) act by resisting sulfate penetration.

The aggregate type is also involved at the level of chemical resistance. It has been assumed that aggregates can react with sulfate solution, since limestone aggregates contain carbonate ions, which is a necessary reactant in thaumasite formation, and degradation of this type has been observed in concrete with limestone aggregates [SAN 01]. However, other studies by the same authors showed a positive effect of limestone aggregates on concrete which contain slag. Some hypotheses have been developed, but the mechanism has not been clearly identified since other effects, such as calcium ion transport and the interface transition zone, can interfere.

Finally, given that degradation propagates through the paste, the paste volume and conversely the aggregate volume are likely to have a significant influence.

Mineral admixtures

Fly ash

Fly ash often noticeably improves sulfate resistance, at least at substitution rates less than 30% (in the absence of data for rates higher than this limit which is set in the standard) [DUV 92, NEV 04].

A study of concrete mixtures containing 0.15 and 30% fly ash, with a constant eq. binder, found a reduction in swelling with increasing fly ash proportions. Porosity and water absorption coefficients were close, but the chloride ion diffusion coefficient reduced with the proportion of fly ash, regardless of the curing mode. In this type of immersion test, diffusion is actually the principal mechanism by which aggressive ions penetrate the concrete, and it is at this level that fly ash seems to have a positive effect.

Fly ash also acts on the chemical resistance of the binder, via the pozzolanic reaction. This consumes the calcium hydroxide (portlandite), which would otherwise react with sulfates and initiate binder degradation (degradation via ettringite and gypsum formation).

Nevertheless, fly ash can have a varied chemical and mineral composition and therefore different levels of effectiveness in increasing concrete resistance to sulfates.

Slag

Standards for cements used in structures exposed to seawater (NF P 15-317) and high sulfate content (XP P 15-319) specify that the proportion of granulated blast furnace slag in CHF-CEM III cements must be at least 60%.

Some studies have been carried out on this subject, and demonstrated that binders with high slag content have good resistance in sulfate rich waters, in tests when the sulfate solution concentration was higher than the threshold of the most severe exposure class (6 g/L in water and 12 g/kg in soil).

A study by Higgins [HIG 03] showed that the resistance of concrete, which has a slag proportion of 60 or 70% in the binder, is clearly better than that of Portland cement concrete – in sodium sulfate solutions – and is comparable to magnesium sulfate solutions, with concentrations in the order of 1.5% (of SO_3 by mass). It is improved with small amounts of calcium carbonate (4%).

Several explanations have been proposed for these kinds of results [DUV 92]. On the one hand, the use of slag reduces the calcium hydroxide and C_3A contents. This can explain the clear improvement in exposure to sodium sulfate, and the more mitigated results obtained in magnesium sulfate attacks. On the other hand, the addition of slag increases hydrate compactness and reduces the average pore size, which causes a reduction in transport properties.

Limestone filler

Taking limestone additives into account in calculating the eq. binder, in a sulfate rich environment, is not possible (from Table NA.F.1 [NFE 04]), and the cement standards only permit the introduction of small amounts (less than 5%) of secondary constituents, such as calcium carbonate (the principal component of limestone), in cements.

Several reasons can explain this position. Limestone fillers limit swelling, but this effect is temporary [DUV 92]. Some results are contradictory and the effect seems to depend on other parameters – the cation and cement types, for example. In any case, these constituents include carbonate ions, which can lead to thaumasite formation if the temperature and humidity conditions are favorable [IRA 03].

This study focused on the effect of limestone additions, at substitution rates of 10 and 20%, on the sulfate resistance of mortars immersed in a sodium sulfate solution for two years (ASTM C 1012 test). The measurements and observations undertaken showed that binders with limestone fillers have a wide range of vulnerability to this type of attack.

Conclusion

With regard to the influence of formulation parameters and the significance of different indicators for sulfate resistance, the sulfate type (cation) must also be included. The proportion of cement in the mixture seems to have, through the resultant compactness, a dominating influence with regard to the C_3A content of the cement. Sulfate resistance reduces when the W/C ratio increases, which appears to be linked to a variation in the diffusion coefficients for different species.

The effects of mineral admixtures are more complicated. Limestone fillers appear to have a detrimental effect, but this trend is reversed at low rates (around 5%) in the presence of slag. Slag has, at proportions higher than 60% in the binder, a favorable effect. In any case, the pozzolanic reaction seems to have unfortunate effects on magnesium sulfate resistance.

4.5. Conclusion

SCC durability is influenced by the same environments and parameters as conventional concrete. For a given environment or *exposure class* [NFE 04], they are subjected to the same prescriptive or performance requirements. Since the phenomena are identical, variations in composition parameters produce the same effects in performance terms, even if the variation ranges are sometimes different.

On the basis of a sound knowledge of degradation mechanisms and influential parameters, it is possible to optimize the BAP composition in the same way as a conventional concrete. Conversely, for identical composition parameters, SCC and conventional concrete will have similar behaviors with regard to durability.

SCC durability is directly linked to fresh state property criteria. The fluidity that is required for self-compacting behavior, which is generally obtained through a high super-plasticizer content, seems to be a good asset since it leads to reductions in the trapped air content and hence improves compactness and strength.

Criteria for SCC when fresh are met by an increased paste volume and increased mineral admixtures content. The increase in paste volume slightly worsens the mechanical properties, but does not seem to have a significant effect on durability (carbonation, chloride ion penetration, leaching and external sulfate attacks).

The choice of mineral admixtures reveals a particular importance, linked to the degradation mechanism to which the concrete is exposed. Limestone fillers seem to be the most used since they allow the fresh state criteria to be met relatively easily. This type of mineral admixture also has a positive effect on diffusion. However, pozzolanic-type additives (fly ash, silica fumes) or latent hydraulic (slag or other industrial by-products) seem to be more effective in reducing diffusion.

Finally, limestone filler increases vulnerability to acid and external sulfate attacks. In order to optimize the composition of SCC, the environment to which it is going to be exposed must be known.

The results presented above concern the potential durability, i.e. the behavior of concrete subjected to characterization and ageing tests in the laboratory. It will be interesting to complement these results with experimental data, which would certainly allow the necessary quality steps for SCC production to be checked against real concrete durability.

Since some formulation parameters such as effective water have a dominating influence on durability, they must be controlled throughout production. Finally, the use of SCC is rarer in structures. However, this usage has been shown to be desirable and not detrimental to durability, even if concrete is exposed to freezing [LAR 06].

4.6. Bibliography

[AFP 97] AFPC-AFREM, "Mode opératoire recommandé, essai de carbonatation accéléré, mesure de l'épaisseur de béton carbonate", *Compte-rendu des journées techniques AFPC-AFREM. Durabilité des bétons, méthodes recommandées pour la mesure des grandeurs associées à la durabilité*, p. 153-158, Toulouse, 11-12 December 1997.

[ALE 99] ALEXANDER M.G., MAGEE B.J., "Durability performance of concrete containing condensed silica fume", *Cement and Concrete Research*, vol. 29, p. 917-922, 1999.

[ASS 04] ASSIÉ S., Durabilité des bétons auto-plaçants, PhD Thesis, INSA Toulouse, 2004.

[AST 00] ASTM C 1012-95a, "Standard test method for length change of hydraulic-cement mortars exposed to a sulfate solution", *Annual Book of ASTM Standards*, vol. 4, 2000.

[ATT 02] ATTIOGBE E.K., SEE H.T., DACZKO J.A., "Engineering properties of self-consolidating concrete", *First North American Conference on the Design and Use of Self-Consolidating Concrete*, p. 371-376, ACBM Center, Chicago, USA, 12-13 November 2002.

[AUD 03a] AUDENAERT K., DE SCHUTTER G., "Influence of moisture on the carbonation of self-compacting concrete", *Proceedings (SP 212) of the Sixth CANMET/ACI International Conference on Durability of Concrete*, p. 451-465, Greece, June 2003.

[AUD 03b] AUDENAERT K., DE SCHUTTER G., "Chloride penetration in self-compacting concrete", *Proceedings of 3rd International RILEM Symposium on Self-Compacting Concrete*, p. 818-825, Reykjavik, Iceland, 17-20 August 2003.

[BAD 06] BADOZ C., FRANCISCO P., ROUGEAU P., "A performance test to estimate durability of concrete products exposed to chemical attacks", *Proceedings of the Second International Congress of FIB*, Naples, Italy, 5-8 June 2006.

[BAR 04] BAROGHEL-BOUNY V. *et al.*, *Conception des bétons pour une durée de vie donnée des ouvrages. Maîtrise de la durabilité vis-à-vis de la corrosion des armatures et de l'alcali-réaction. Etat de l'art et guide pour la mise en œuvre d'une approche performantielle et prédictive sur la base d'indicateurs de durabilité*, Association Française de Génie Civil, 2004.

[BAR 96] BARON J., GAGNE R., OLLIVIER J.P., "Viser la durabilité", in J. BARON, J.P. OLLIVIER (eds), *Les Bétons, Bases et données pour leur formulation*, Eyrolles, Paris, 1996.

[BAS 01] BASHEER L., KROPP J., CLELAND D.J., "Assessment of the durability of concrete from its permeation properties: a review", *Construction and Building Materials*, vol. 15, p. 93-103, 2001.

[BOU 94] BOURDETTE B., Durabilité du mortier: prise en compte des auréoles de transition dans la caractérisation et la modélisation des processus physiques et chimiques d'altération, PhD Thesis, INSA Toulouse, 1994.

[BUR 06] BURDEN D., The durability of concrete containing high levels of fly ash, Thesis, University of New Brunswick, 2003, PCA R&D Serial n°2989, Portland Cement Association, 2006.

[CAR 96] CARDE C., FRANÇOIS R., TORRENTI J.M., "Leaching of both calcium hydroxide and C-S-H from cement paste: modelling the mechanical behaviour", *Cement and Concrete Research*, vol. 26, p. 1257-1268, 1996.

[CHA 97] CHAUSSADENT T., "Analyse des mécanismes de carbonatation du béton", *Compte-rendu des journées techniques AFPC-AFREM. Durabilité des bétons, méthodes recommandées pour la mesure des grandeurs associées à la durabilité*, p. 75-87, 11-12 December 1997.

[COQ 99] COQUILLAT G., Recherche collective Bétons avec additions, Béton B 25 – Type "Bâtiment, CEBTP report, 1999.

[DER 94] DERACHE C., GUIRAUD P., PLUMAT M., ROUGEAU P., THOMAS B., VALLÈS M., VICHOT A., "La durabilité des bétons", *Collection technique CIMbéton*, vol. 48, 1994.

[DES 03] DE SCHUTTER G., AUDENAERT K., BOEL V., VANDEWALLE L. *et al.*, "Transport properties in self-compacting concrete and relation with durability: overview of a Belgian research project", *Proceedings of 3rd International RILEM Symposium on Self-Compacting Concrete*, p. 799-807, Reykjavik, Iceland, 17-20 August 2003.

[DES 04] DE SCHUTTER G., AUDENAERT K., "Evaluation of water absorption of concrete as a measure for resistance against carbonation and chloride migration", *Materials and Structures*, vol. 37, p. 591-596, 2004.

[DUV 92] DUVAL R., HORNAIN H., "La durabilité du béton vis-à-vis des eaux agressives", in J. BARON and J.P. OLLIVIER (eds.), *La durabilité des bétons*, Presses de l'ENPC, Paris, 1992.

[FRA 01] FRANÇOIS R., FRANCY O., CARÉ S., BAROGHEL-BOUNY V., LOVERA P., RICHET C., "Mesure du coefficient de diffusion des chlorures, Comparaison entre régime permanent et régime transitoire", *Revue française de génie civil*, vol. 5, no. 2-3/2001, p. 309-329, 2001.

[GAL 05] GALLÉ C. *et al.*, "Intermediate long-lived nuclear waste management: an integrated approach to assess the long-term behaviour of cement-based materials in the context of deep disposal", *SRNL Workshop on Cementitious Materials*, Aiken, United States, 2005.

[GRA 07] ARLIGUIE G., HORNAIN H., *GranDuBé, Grandeurs associées à la durabilité des bétons*, Presses de l'Ecole Nationale des Ponts et Chaussées, Paris, 2007.

[GUI 04] GUILLON E., Durabilité des matériaux cimentaires, modélisation de l'influence des équilibres physico-chimiques sur la microstructure et les propriétés mécaniques résiduelles, PhD Thesis, Ecole Normale Supérieure de Cachan, 2004.

[HIG 03] HIGGINS D.D., "Increased sulfate resistance of ggbs concrete in the presence of carbonate", *Cement and Concrete Composites*, vol. 25, p. 913-919, 2003.

[IRA 03] IRASSAR E.F., BONAVETTI V.L., GONZALEZ M., "Microstructural study of sulphate attack on ordinary and limestone Portland cements at ambient temperature", *Cement and Concrete Research*, vol. 33, p. 31-41, 2003.

[KAM 03] KAMALI S., GÉRARD B., MORANVILLE M., "Modelling the leaching kinetics of cement-based materials – influence of materials and environment", *Cement and Concrete Composites*, vol. 25, p. 451-458, 2003.

[KHO 07] KHOKHAR M.I.A., ROZIÈRE E., GRONDIN F., LOUKILI A., "Effect of mineral additives on some of durability parameters of concrete", *International Conference on Advances in Cement Based Materials and Applications to Civil Infrastructure (ACBM-ACI)*, Lahore, Pakistan, 12-14 December 2007.

[KLI 41] KLINKENBERG L.J., "The permeability of porous media to liquid and gases", *Drilling and Production Practice*, American Petroleum Institute, p. 200-213, 1941.

[KOL 04] KOLIAS S., GEORGIOU C., "The effect of paste volume and of water content on the strength and water absorption of concrete", *Cement and Concrete Composites*, vol. 27, no. 2, February 2005, p. 211-216, 2004.

[LAR 99] DE LARRARD F., "Concrete mixture-proportioning – A scientific approach", *Modern Concrete Technology Series*, no. 9, E&FN SPON, 1999.

[LAR 06] LARIVE C., FÉRON C., DIERKENS M., DROY P., DUTILLOY P.H. "Une voûte de tunnel en béton auto-plaçant", *Journées des Sciences de l'Ingénieur*, Marne la Vallée, 5-6 December 2006.

[LEB 01] LE BELLEGO C., Couplage chimie mécanique dans les structures en béton armé attaquées par l'eau – Etude expérimentale et analyse numérique, PhD Thesis, Ecole Normale Supérieure de Cachan, 2001.

[MAI 00] MAINGUY M., TOGNAZZI C., TORRENTI J.M., ADENOT F., "Modelling of leaching in pure cement paste and mortar", *Cement and Concrete Research*, vol. 30, p. 83-90, 2000.

[MEH 79] MEHTA P.K., PIRTZ D., POLIVKA M., "Properties of alite cements", *Cement and Concrete Research*, vol. 9, p. 439-450, 1979.

[MOO 05] MOON H.Y., KIM H.S., CHOI D.S., "Relationship between average pore diameter and chloride diffusivity in various concretes", *Construction and Building Materials*, vol. 20, no. 9, p. 725-732, 2005.

[MOR 04] MORANVILLE M., KAMALI S., GUILLON E., "Physicochemical equilibria of cement-based materials in aggressive environments – experiment and modeling", *Cement and Concrete Research*, vol. 34, p. 1569-1578, 2004.

[MOR 01] MORTSELL E., RODUM E., "Mechanical and durability aspects of SCC for road structures", *Proceedings of 2nd International Symposium on Self-Compacting Concrete*, p. 459-468, Tokyo, Japan, 23-25 October 2001.

[NEV 04] NEVILLE A., "The confused world of sulfate attack on concrete, Review", *Cement and Concrete Research*, vol. 34, p. 1275-1296, 2004.

[NFE 04] NF EN 206-1, Béton – Partie 1: Spécification, performances, production et conformité, AFNOR, April 2004.

[NFP 95a] NF P 15-317, Liants hydrauliques – Ciments pour travaux à la mer, Recueil de normes françaises, Béton et constituants du béton, vol. 3, AFNOR, 1995.

[NFP 95b] NF P 15-319, Liants hydrauliques – Ciments pour travaux en eaux à hautes teneurs en sulfates, Recueil de normes françaises, Béton et constituants du béton, vol. 3, AFNOR, 1995.

[NGU 06] NGUYEN V.H., NEDJAR B., COLINA H., TORRENTI J.M., "A separation of scales analysis for the modelling of calcium leaching in concrete", *Computer Methods in Applied Mechanics and Engineering*, vol. 195, p. 7196-7210, 2006.

[NTB 00] NT Build 443, Essai d'immersion pour la détermination du coefficient de diffusion des chlorures en régime non stationnaire par mesure du profil de penetration, Nordtest Method, 2000.

[NTB 99] NT Build 492, Concrete, mortar and cement-based repair materials: Chloride migration coefficient from non-steady-state migration experiments, Nordtest Method, 1999.

[PAP 91] PAPADAKIS V.G., VAYENAS C.G., FARDIS M.N., "Fundamental modeling and experimental investigation of concrete carbonation", *ACI Materials Journal*, vol. 88, no. 4, July-August 1991.

[PAP 00] PAPADAKIS V.G., "Effect of supplementary cementing materials on concrete resistance against carbonation and chloride ingress", *Cement and Concrete Research*, vol. 30, p. 291-299, 2000.

[PER 99] PERRATON D., AITCIN P.C., CARLES-GIBERGUES A., "Mesure de la perméabilité aux gaz des bétons: perméabilité apparente et perméabilité intrinsèque", *Bulletin des Laboratoires des Ponts et Chaussées*, vol. 221, p. 69-78, 1999.

[PIC 01] PICANDET V., KHELIDJ A., BASTIAN G., "Effect of axial compressive damage on gas permeability of ordinary and high-performance concrete", *Cement and Concrete Research*, vol. 31, p. 1525-1, 2001.

[RIL 99] RILEM TC 116-PCD, "Recommendation of TC 116-PCD: Tests for gas permeability of concrete – Preconditioning of concrete test specimens for the measurement of gas permeability and capillary absorption of water – Measurement of the gas permeability of concrete by the RILEM – CEMBUREAU method – Determination of the capillary absorption of water of hardened concrete", *Materials and Structures*, vol. 32, no. 217, 1999.

[PLA 06] PLANEL D., SERCOMBE J., LE BESCOP P., ADENOT F., TORRENTI J.M., "Long-term performance of cement paste during combined calcium leaching-sulfate attack: kinetics and size effect", *Cement and Concrete Research*, vol. 36, p. 137-143, 2006.

[ROZ 07] ROZIÈRE E., GRANGER S., TURCRY P., LOUKILI A., "Influence of paste volume on shrinkage cracking and fracture properties of self-compacting concrete", *Cement and Concrete Composites*, vol. 29, no. 8, p. 626-636, September 2007.

[ROZ 07] ROZIÈRE E., Etude de la durabilité des bétons par une approche performantielle, PhD Thesis, Ecole Centrale de Nantes, 2007.

[ROZ 09] ROZIÈRE E., LOUKILI A., CUSSIGH F., "A performance based approach for durability of concrete exposed to carbonation", *Construction and Building Materials*, vol. 23, no. 1, p. 190-199, January 2009.

[ROZ 09] ROZIÈRE E., LOUKILI A., EL HACHEM R., GRONDIN F., "Durability of concrete exposed to leaching and external sulphate attacks", *Cement and Concrete Research*, vol. 39, no. 12, p. 1188-1198, December 2009.

[ROZ 11] ROZIÈRE E., LOUKILI A., "Performance-based assessment of concrete resistance to leaching", *Cement and Concrete Composites*, vol. 33, no. 4, p. 451-456, April 2011.

[SAM 07] SAHMARAN M., KASAP O., DURU K., YAMAN I.O., "Effects of mix composition and water-cement ratio on the sulfate resistance of blended cements", *Cement and Concrete Composites*, vol. 29, p. 159-167, 2007.

[SAN 01] SANTHANAM M., COHEN M.D., OLEK J., "Sulfate attack research – whither now?", *Cement and Concrete Research*, vol. 31, p. 845-851, 2001.

[STO 79] STOCK A.F., HANNANT D.J., WILLIAMS R.I.T., "The effect of aggregate concentration upon the strength and modulus of elasticity of concrete", *Mag. Conc. Res.*, no. 109, p. 225-234, 1979, in A.M. NEVILLE, *Propriétés des bétons*, Eyrolles, Paris, 2000.

[TAN 96] TANG L., Chloride transport in concrete – measurement and prediction, PhD Thesis, Publication P-96:6, Dept. of Building Materials, Chalmers University of Technology, Gothenburg, Sweden, 1996.

[TAN 01] TANGE JEPSEN M., MUNCH-PETERSEN C., BAGER D., "Durability of resource saving "green" types of concrete", *Proceedings FIB-symposium "Concrete and environment"*, Berlin, Germany, October 2001.

[THI 06] THIERRY M., "Modélisation de la carbonatation atmosphérique des matériaux cimentaires, Prise en compte des effets cinétiques et des modifications microstructurales et hydriques", *Etudes et recherches des laboratoires des Ponts et Chaussées*, OA 52, 2006.

[THO 99] THOMAS M.D.A., SHEHATA M.H., SHASHIPRAKASH S.G., HOPKINS D.S., CAIL K., "Use of ternary cementitious systems constaining silica fume and fly ash in concrete", *Cement and Concrete Research*, vol. 29, p. 1207-1214, 1999.

[TRA 99] TRÄGARDH J., "Microstructural features and related properties of self-compacting concrete", *Proceedings of First International RILEM Symposium on Self-Compacting Concrete*, p. 175-186, Stockholm, Sweden, 13-15 September 1999.

[TRA 03] TRÄGARDH J., SKOGLUND P., WESTERHOLM M., "Frost resistance, chloride transport and related microstructure of field self-compacting concrete", *Proceedings of 3rd International RILEM Symposium on Self-Compacting Concrete*, p. 881-891, Reykjavik, Iceland, 17-20 August 2003.

[SAK 98] SAKATA K., "Durability of self-compacting concrete and low heat high performance concrete", *CONSEC 98 International Conference*, p. 2057-2064, Tromso, Norway, 1998.

[TIA 00] TIAN B., COHEN M.D., "Does gypsum formation during sulfate attack on concrete lead to expansion?", *Cement and Concrete Research*, vol. 30, p. 117-123, 2000.

[TRU 00] TRUC O., OLLIVIER O., CARCASSÈS M., "A new way for determining the chloride diffusion coefficient in concrete from steady state migration", *Cement and Concrete Research*, vol. 30, p. 217-226, 2000.

[TUR 04] TURCRY P., Retrait et fissuration des bétons auto-plaçants, influence de la formulation, PhD Thesis, Ecole Centrale de Nantes and University of Nantes, 2004.

[VEZ 99] VEZOLE P., Recherche collective Bétons avec additions, synthèse, report from CEBTP, 1999.

[ZHU 01] ZHU W., QUINN J., BARTOS P.J.M., "Transport properties and durability of self-compacting concrete", *Proceedings of 2nd International Symposium on Self-Compacting Concrete*, p. 451-458, Tokyo, Japan, 23-25 October 2001.

[ZHU 01] ZHU W., BARTOS P.J.M., "Permeation properties of self-compacting concrete", *Cement and Concrete Research*, vol. 33, p. 921-926, 2001.

Chapter 5

High Temperature Behavior of Self-Compacting Concretes

5.1. Introduction

In order to obtain the desired fluidity and stability characteristics in the fresh state, SCC composition is significantly different to that of conventional concretes (Chapter 1). This composition is characterized first of all by higher chemical admixture proportions (high range water reducing admixtures and viscosity enhancing admixtures). But SCCs also contain higher paste volumes than conventional concretes (CC). For economic reasons, the large paste volume is commonly achieved by adding supplementary cementitious materials (SCM) to the cement, particularly fillers. Furthermore, the granular skeleton in SCCs is markedly different to that of conventional concretes. The maximum grain diameter is often reduced relative to that of conventional concretes and the CA/S (coarse

---

Chapter written by Hana FARES, Sébastien RÉMOND, Albert NOUMOWÉ and Geert DE SCHUTTER.

aggregate/sand) ratio is usually close to 1 while in conventional concretes it is generally greater than 1.5. These particular composition details are likely to significantly modify the mechanical behavior of SCCs, particularly their high temperature behavior. This chapter presents state of the art understanding of high temperature SCC behavior.

First, the physico-chemical changes which occur inside SCCs when a high temperature exposure is examined, from the perspective of knowledge acquired on high temperature behavior of conventional vibrated concretes.

Next, the evolution of physical properties, such as porosity and permeability, and mechanical properties (compressive and tensile strength, and elastic modulus) when a rise in temperature is studied and compared with data on conventional concretes.

Finally, the stability of SCCs at high temperatures and the effectiveness of polypropylene fibers with regard to explosive spalling are presented.

5.2. Changes in SCC microstructure and physico-chemical properties with temperature

5.2.1. *Physico-chemical properties*

During heating, cement paste and aggregates are subjected to physico-chemical transformations as a function of temperature, such as phase changes, shrinkage or expansion, cracking, etc., which significantly alter the physical and mechanical properties of the material. Research into these alterations at the microstructural level is indispensible for understanding the mechanisms responsible for the degradation of the material's macroscopic properties with temperature.

Chemical and mineralogical changes

Thermal analysis (DTA, differential thermal analysis, and TGA, thermal gravimetric analysis) allow the evolution of the matrix composition to be studied as the temperature increases.

1m³	SCC 25	SCC 40	CC 40
CEM I 52.5 N	-	350	373
CEM II 32.5 R	328	-	-
Limestone filler	225	130	-
Sand 0/4	795	857	913
Aggregate 4/22.5	745	742	790
Effective water	199	200	202
Slump test (cm)	65	71	19
Paste volume (liters)	393.4	361.8	321.2
W/Binder	0.52	0.52	0.54
W/C	0.61	0.57	0.54
W/Powder	0.36	0.42	0.54
Filler/Binder	0.59	0.34	0
A/S	0.94	0.87	0.87
$M_{water}/M_{concrete}$ (%)	8.86	8.78	8.87
Real strength at 90 days (MPa)	37	54	41

Table 5.1. *Formulations for SCCs and CC used in Fares et al. [FAR 09, FAR 10]*

Fares *et al.* [FAR 10] carried out DTA-TGA on two SCCs and an CC (see Table 5.1). These tests were conducted on mortar from tested concretes whilst avoiding taking aggregates and were concerned with both concretes that had not been previously subjected to any thermal treatment and concretes degraded after thermal treatments at 150, 300, 450 and 600°C (slow heating at 1°C/min).

For unheated concretes (Figure 5.1), the 6 endothermic peaks present are associated with changes in the crystalline state of dehydration of certain hydrates in the cement paste.

The double peak at 110 and 130°C is attributed to (absorbed) water loss from particular hydrates: C-S-H and ettringite [PER 86]. At 200°C, the presence of a peak indicates dehydration of hydrated calcium mono-carboaluminate [NON 99].

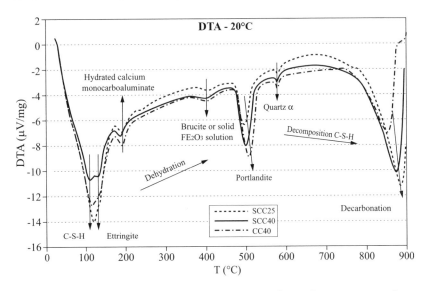

Figure 5.1. *Differential thermal analysis on unheated concrete samples (SCC and CC) [FAR 10]*

Between 200 and 300°C, Khoury [KHO 95], Noumowé [NOU 95] and Richard [RIC 99] attribute slight flux variations to the continuing C-S-H dehydration.

At 400°C, a small peak is observed whose identity cannot be clearly established. A similar transformation has been observed by Sha *et al.* [SHA 99] for cement pastes. Those authors attribute this peak to a change in the crystalline state or to dehydration of a solid solution of Fe_2O_3 [SHA 99]. However, other sources [PER 86] identified this peak as brucite decomposition ($Mg(OH)_2$).

Between 450 and 550°C, the peak which corresponds to portlandite decomposition to free lime [NOU 95, PLA 02] is observed (dehydroxylation).

At 573°C, the allotropic transformation from quartz-α to quartz-β is observed, accompanied by an expansion in size (responsible for cracking siliceous aggregates) [PLA 02].

Between 600 and 700°C, C-S-H decomposes and transforms to a new bicalcium form of silicate (β-C_2S) [PLA 02]. From 700 to 900°C, the peak is attributed to the decomposition of limestone aggregates and fillers. This is accompanied by a release of CO_2 [NOU 95, KHO 95].

Figure 5.2 shows the DTA curves obtained with the same materials having been subjected to the thermal treatments described above [FAR 10]. In comparison with Figure 5.1, several peaks have a reduced intensity, or are almost absent, which corresponds to the partial, or total, disappearance, or particular phases during the different thermal treatments.

After treatment at 150°C, the peak at 110-130°C reduces as a result of the loss of free water, at the start of the first phase of C-S-H and ettringite dehydration [NOU 95]. For the other constituents, no other change is observed.

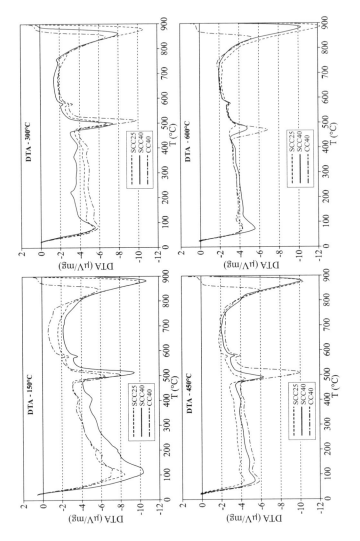

Figure 5.2. *Thermo-differential analysis of concrete samples (SCC and CC) subjected to different heat treatments [FAR 10]*

Following heating to 300°C, the peaks which corresponds to the loss of free water, water bonded to C-S-H (first decomposition phase) or to ettringite such as those bonded to hydrate calcium mono-carboaluminate is eliminated.

After heating to 450°C, a net decrease in the peak at 400°C is observed. Therefore, brucite or the solid solution of Fe_2O_3 has disappeared.

For a cycle at 600°C, the peak of dehydroxylation (portlandite at 500°C) is significantly reduced. On the other hand, the peak at 573°C, which corresponds to the allotropic transformation of quartz-α to quartz-β, stays practically unchanged. Since this transformation is reversible, quartz-α reforms after cooling.

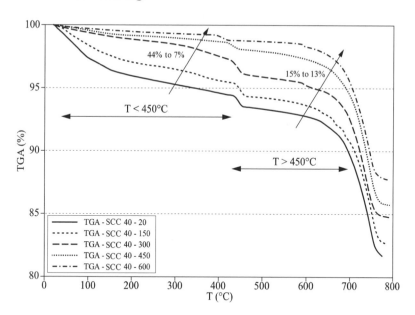

Figure 5.3. *Thermogravimetric analysis of SCC 40 subjected to different heat treatments [FAR 10]*

During heating, water is progressively eliminated from the concrete as a function of its bond energy (first of all free water, then bonded water). In the same way, hydrates and aggregates undergo transformations according to the types of minerals that they are made of. The curves in Figures 5.1 and 5.2 show significant differences in the materials with regard to the heights of different peaks. The compositions of the three concretes studied differ in the type of cement used, the presence of fillers (in SCC) and by their proportions of different constituents. The quantities of hydrates formed from one concrete to another are therefore naturally different, which explains the differences between the peak heights. On the other hand, the positions of the peaks are the same overall for all the concretes; the types of transformations which occur inside the material are also therefore the same for all the concretes.

Figure 5.3 shows the change in mass obtained via thermogravimetric analysis (TGA) of an SCC40 concrete sample before and after heat treatment at 150°C, 300°C, 450°C and 600°C [FAR 10]. These measurements were taken during heating at 20°C/min from 20°C to 800°C.

Two important mass losses occur at 450°C and 700°C and correspond to decomposition phenomena (portlandite dehydroxylation and calcite decarbonation) [NOU 95, PLA 02, KHO 95]. Part of the graph with low gradients show information on water leaving the concrete [KHO 95, NOU 95, RIC 99, NON 99, SHA 99, PER 86]. According to analysis by Divet *et al.* [DIV 05], between 20°C and 450°C, water is considered to be weakly bonded to cement hydrates and, in particular, to hydrate calcium silicates (C-S-H) whereas between 450°C and 700°C, it is considered to be strongly bonded to hydrates (C-S-H).

Applying a heat treatment to concrete does not significantly change the appearance of the TGA curve

(Figure 5.3) but it does reduce the mass loss measured during the test. For a concrete that has not been heat treated, the mass loss between 20°C and 450°C represents 44% of the total mass loss, where it is no more than 7% for concrete which has been heated up to 600°C. Exposure to high temperature therefore leads first of all to loss of weakly bonded water which is, moreover, accompanied by shrinkage which contributes to the first appearance of cracks in the concrete.

The mass loss which occurs between 450°C and 700°C is smaller than that measured between 20°C and 450°C. This mass loss represents 15% of the total mass loss for a concrete which has not been heat treated. Furthermore, applying a 600°C heat treatment only leads to a moderate reduction in this mass loss (13% of the total mass loss).

Between 450°C and 700°C, phase changes such as the transformation of C-S-H to bicalcium silicates (β-C$_2$S) and portlandite dehydroxlyation occur and lead to the loss of water which was chemically bound to the hydrates. This exacerbates the cracks which were already observed.

Thermo-gravimetric and differential thermal analysis (10°C/min to 1,200°C) have also been carried out by [LIU 06, YE 07] on different paste types which have not undergone any heat treatments (HPCP, TCP and SCCP, Figure 5.4).

The TGA and DTA curves have similar shapes for the three cement paste types up to 600°C. On the other hand, a significant peak appears between 700°C and 800°C in the DTA curves for the two SCC pastes (SCCP1 and SCCP2, Figure 5.4). These curves are linked to a significant mass reduction (TGA). The SCCs tests contain limestone fillers. This mass loss therefore comes about from decarbonation of these limestone fillers according to the reaction $CaCO_3 \rightarrow CaO + CO_2$.

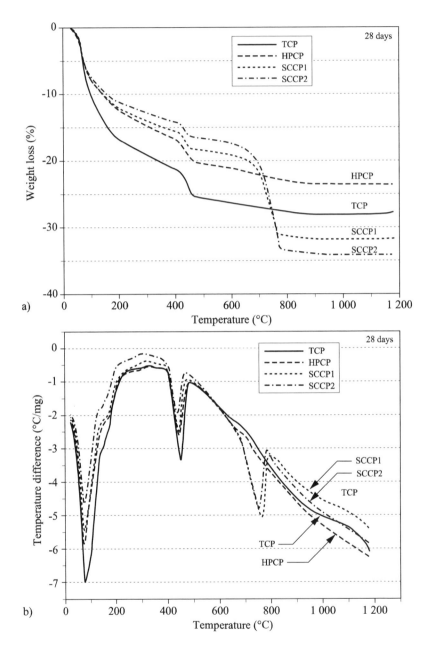

Figure 5.4. *TGA and DTA on three cement paste types (HPCP, TCP and SCCP) [YE07]*

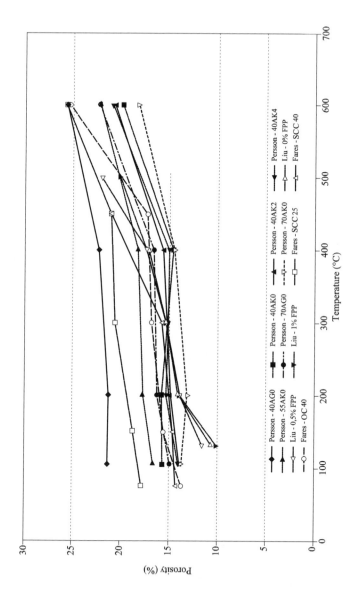

Figure 5.5. *Change in porosity with temperature (Mercury porosity: Liu et al. – Water porosity: Persson and Fares et al.) [PER 04, LIU 06, FAR 09]*

Porosity

Transformations inside the cement paste, particularly hydrate decomposition and differential deformations between paste and aggregates, lead to an increase in concrete porosity which is detrimental to the material's mechanical behavior. The increase in porosity during heating has been widely studied for different types of concretes. Figure 5.5 shows the change in porosity with temperature for different SCCs and some CCs obtained by several authors [PER 04, LIU 06, FAR 09].

Overall, porosity increases linearly with temperature, regardless of the type of concrete studied. For several related concretes in different studies, acceleration in the increase in porosity is observed when the temperature reaches 400°C – 450°C. This acceleration corresponds in part to an increase in cracking in the material, notably at the paste-aggregate interface.

The physico-chemical transformations mentioned above, in particular the loss of chemically bonded water, results in a significant increase in concrete porosity, which leads to a change in the mechanical properties of the material such as its transfer properties. These occur for all concretes tested and can therefore be generalized.

Figure 5.6 shows SEM micrographs of SCC samples (SCC 40 from work by Fares *et al.* [FAR 10] cited in the previous section) at ambient temperature and after exposure to 150°C, 300°C, 450°C and 600°C. Up to 150°C, the surface of concrete does not show any changes in facies. No cracks can be seen. At 300°C, a few rare cracks appear, notably at the paste-aggregate interface. Cracking is clearly more pronounced in samples heated to 450°C and especially to 600°C. At these temperatures, cracks are observed at the paste-aggregate interface but also right inside the paste and aggregates, particularly at 600°C. Cross-grain cracks observed at this temperature are essentially the result of quartz (SiO_2) in the aggregates. At 573°C, the allotropic transformation from quartz-α to quartz-β takes place. This reversible transformation has important consequences for the physical properties of quartz and in particular causes expansion (0.8% by volume).

Some studies have also focused on changes in porosity with temperature of SCCs containing polypropylene fibers (PPF). The presence of fibers in concrete allows significant improvements in the material's thermal stability (see section 5.4).

Persson [PER 04] studied the changes in porosity of different SCCs (Table 5.2) subjected to the following temperatures: 105°C, 200°C, 400°C and 600°C.

SCC mix proportions					Temperature (°C) and Porosity (%)			
Name	Cement	W/B	PPF (kg m^{-3})	Filler	105°C	200°C	400°C	600°C
40AG0	CEM I 42.5	0.40	0	Glass	21.3	21.3	22.3	25.6
40AK0	CEM I 42.5	0.40	0	Limestone	15.7	15.7	14.7	19.8
40AK2	CEM I 42.5	0.40	2	Limestone	14.0	15.0	15.6	20.7
40AK4	CEM II 42.5	0.40	4	Limestone	14.1	15.3	15.0	21.0
55AK0	CEM II 42.5	0.55	0	Limestone	16.7	17.8	18.3	22.2
70AG0	CEM I 42.5	0.40	0	Glass	15.0	16.3	16.7	22.3
70AK0	CEM I 42.5	0.40	0	Limestone	13.8	13.2	14.6	18.3

Table 5.2. *Changes in porosity with increasing temperature [PER 04]*

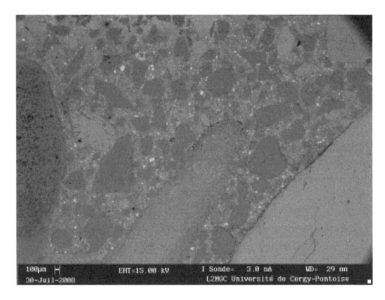

20°C

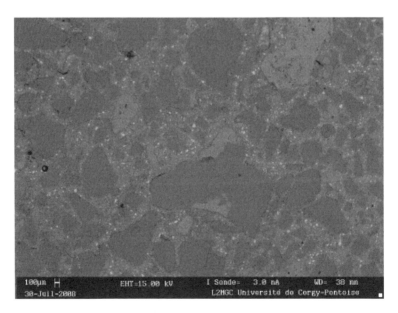

Figure 5.6. (continued) *150°C*

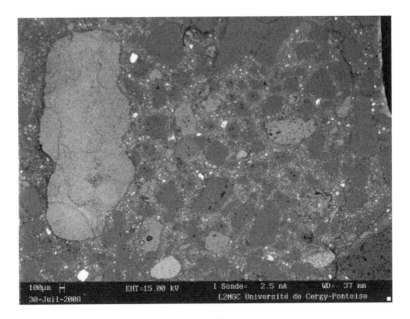

300°C

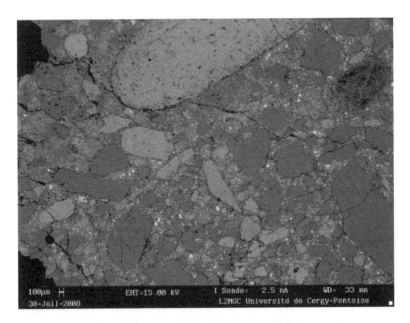

Figure 5.6. (continued) *450°C*

600°C

Figure 5.6. *Micrograph of SCC 40 after exposure to different temperatures (approximate dimensions 225 x 170 µm) [FAR 10]*

As we have seen already (Figure 5.5) a large increase in porosity is observed between 400°C and 600°C for all concretes. This increase is probably due to the transformation of quartzites and water loss from cement hydrates during dehydroxylation of portlandite. The increase in cracking in the material also contributes to the increase in porosity, particularly at the paste-aggregate interface. However, no significant increase in porosity is observed in concretes which contain polypropylene fibers when compared to concretes that do not contain fibers.

Nevertheless, we will see in the following section that PPFs contribute towards a significant improvement in thermal stability by increasing concrete permeability in a range of temperatures which corresponds to the range in which concrete explosive spalling can occur.

Liu *et al.* [LIU 08] have also studied the change in porosity with temperature in SCC pastes and HPC pastes which contain different mixture proportions of PPF (0, 0.5 and 1 kg.m⁻³). They compared, using mercury porosimetry tests (during increasing and falling pressure), the size distribution of pores in the different pastes subjected to heating at a rate of 10°C/min up to 500°C.

SCC	Porosity (%)				
	130°C	200°C	300°C	400°C	500°C
0 kg m⁻³ PPF	10.7	14.1	15.5	17.2	-
0.5 kg m⁻³ PPF	11.6	13.9	15.3	17.4	22.1
1 kg m⁻³ PPF	10.2	14.2	15.5	17.3	20.2

Table 5.3. *Change in paste porosity with temperature for SCCs with different PPF contents [LIU 08]*

The porosity results obtained by Liu *et al.* on SCC pastes [LIU08] are shown in Table 5.3. The authors observed a continuous increase in porosity between 130°C and 500°C for all cement pastes, but no significant difference caused by the PPF additive was observed. PPF fusion (at around 170°C) is expected to lead to an increase in paste porosity between 130°C and 200°C relative to a paste without fibers. The low PPF volume in the pastes tested (between 0.75% and 1.5%) undoubtedly explains the observed lack of difference between cement pastes with and without fibers.

On the other hand, as the temperature rises, (between 130°C and 400°C), Liu *et al.* [LIU 08] noticed that the pore distribution did not vary significantly, but the volume of some pores increased.

Permeability

Very few studies [FAR 09, LIU 08] until now have focused on variations in SCC permeability with temperature.

Fares *et al.* [FAR 09] studied the permeability of 2 SCCs and one CC, and showed that permeability increases continuously from 20°C to 600°C. These results were compared, in Figure 5.7, to results from Kanema [KAN 07] which focused on the change in gas permeability of conventional concretes with strengths between 38 and 72 MPa and tested in the same conditions.

Overall, the increases in permeability in SCCs and CCs are comparable. The change in permeability is due to the appearance of a class of pores which modifies the connectivity of the pore network with, in particular, the widening of capillary pores between 75 and 300°C [TSI 97]. According to Gallé *et al.* [GAL 01], these changes are due to the loss of water from the pore network, loss of adsorbed water and cement hydrate decomposition. These phenomena contribute to an increase in size of the capillary pores and crack creation. At temperatures higher than 300°C, this change is the result, not of a change in capillary porosity, but of deterioration in the cement matrix which leads to a change in the concrete's fine porosity [TSI 97] or to the appearance of microcracks [LIU 08, GAL 01].

Liu *et al.* [LIU 08] studied gas permeability of SCC and HPC cement pastes with and without polypropylene fibers (Figure 5.7, Table 5.4).

SCC	Permeability (m^2 x 10^{-15})					
	105°C	130°C	200°C	300°C	400°C	500°C
0% PPF	0.5	0.4	0.2	0.6	2.2	-
0.5% PPF	0.5	0.4	1.4	2.4	2.8	3.0
1% PPF	0.4	0.4	2.4	3.6	4.2	5.2

Table 5.4. *Change in permeability with temperature of SCCs with different PPF contents [LIU 08]*

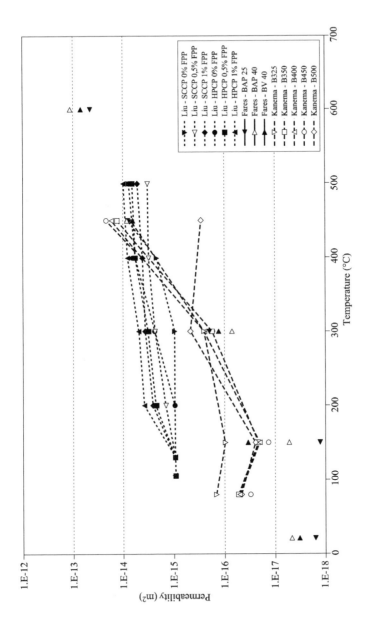

Figure 5.7. *Change in gas permeability with temperature [LIU 08, FAR 09, KAN 07]*

The permeability results are identical for all of the pastes tested when the temperature is below 130°C. However, above this temperature, permeability increases significantly with fiber content. Fusing of the fibers (at around 170°C) results in a very limited increase in porosity which takes into account the low quantities of fibers incorporated in the pastes as seen above.

In any case, the pores created are fine and elongated and they increase the pore network connectivity and therefore the permeability of the material. The higher the fiber content, the more the permeability increases. On the other hand, for SCCs without PPF, the change in permeability only begins at 300°C and is essentially linked to the increase in porosity and connectivity due to dehydration and cracking in the paste.

Figure 5.8 shows the influence of fiber content permeability [LIU 08]. The normalized permeability (relative to concretes without fibers) increases significantly at 200°C with increased fiber content. The influence of PPF is more marked for SCC than for HPC.

The effect of adding 0.5 kg/m^{-3} PPF to SCC is similar to that of adding 1 kg/m^{-3} PPF to HPC. Above 200°C, the normalized permeability tends to decrease. At 400°C, the same permeability level is reached for all formulations tested.

Polypropylene fibers therefore have a significant effect on concrete permeability principally in a temperature range from 105°C to 300°C. Beyond 300°C, the degradation state becomes the same for all of the concretes. The effect of fibers between 105°C and 300°C is beneficial since fiber fusion results in an increase in permeability and therefore a reduction in the risk of explosive spalling which often happens in this temperature range.

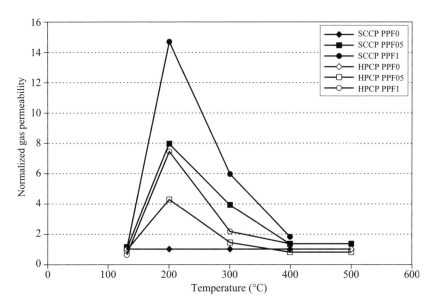

Figure 5.8. *Normalized permeability for SCCP and HPCP relative to concretes without PPF [LIU 08]*

Thermal properties

Jansson [JAN 04] measured the thermal conductivity with temperature rise (Figure 5.9) for 3 types of concretes made with calcium-silica aggregates: conventional concrete (W/C = 0.70; f_{c28} = 38.5 MPa; paste volume = 280 liters), HPC (W/C = 0.28; f_{c28} = 114.2 MPa; paste volume = 254 liters) and SCC (W/C = 0.38; f_{c28} = 92.3 MPa; paste volume = 351 liters). To carry out these tests, the author used different probe types which were selected principally according to the sample size and temperature range.

These tests were carried out up to 600°C. They showed that the thermal conductivity of conventional concretes is inferior to that of HPCs and SCCs (Figure 5.9). On the other hand, HPCs and SCCs have very close thermal conductivity values. The SCCs tested were all high performance SCCs (HP SCC) which may explain this result. Furthermore, no significant difference was observed for the specific heat and thermal diffusivity (Figure 5.10) which seems to be independent of concrete type [JAN 04].

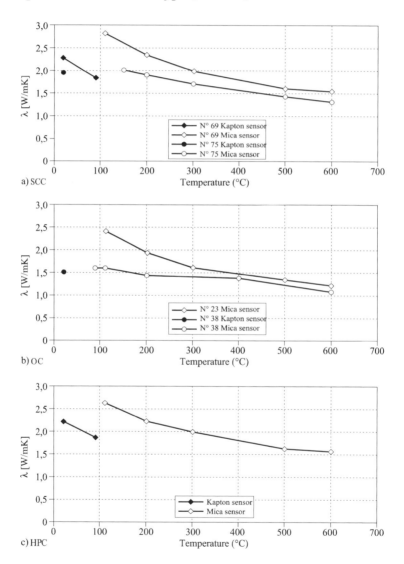

Figure 5.9. *Thermal conductivity of concretes [JAN 04]*

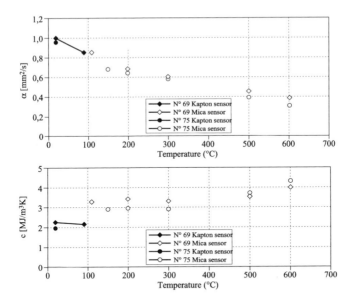

Figure 5.10a. *Thermal diffusivity (top) and specific heat, (bottom) [JAN 04] for SCC*

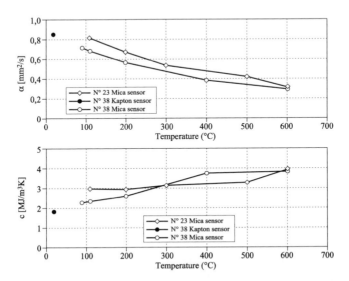

Figure 5.10b. *Thermal diffusivity (top) and specific heat, (bottom) [JAN 04] for CC*

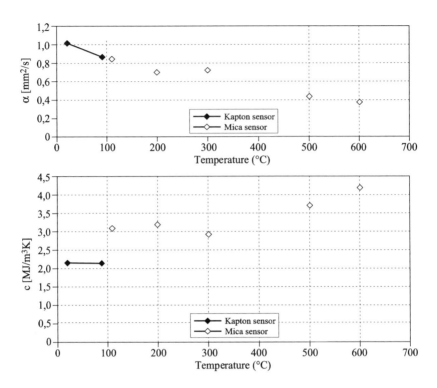

Figure 5.10c. *Thermal diffusivity (a) and specific heat,*
(b) [JAN 04] for HPC

Pineaud [PIN 07] studied thermal expansion of 4 SCCs (M40, M60, M90 and M110, with paste volumes of 323, 312, 406 and 426 liters and W/C ratios of 0.43, 0.38, 0.30 and 0.24 respectively, with strengths from 40 to 110 MPa (Figure 5.11). the M40 and M60 concretes contained limestone fillers whilst M90 contained siliceous filler and M110 silica fume. The tests were carried out at temperatures of 20°C, 120°C, 250°C, 400°C and 600°C.

Pineaud noticed that thermal deformation was similar in the 4 concretes up to 300°C. Beyond 300°C, the M90 and M110 concretes deformed less rapidly than the M40 and M60 concretes which have coincident curves. The measured thermal deformations for the concretes result from deformations in the matrix and aggregates during heating. In general, deformation is essentially governed by expansion of the aggregates [PIM 01, HAG 04]. On the other hand, for M 90 and M 110 concretes, the particular behavior observed above 300°C can be partly attributed to the higher paste volumes in these concretes.

Several studies [JUM 89, KHO 95, HAG 04] have shown this behavior which is particular to the paste: it expands up to around 150°C, then undergoes a significant shrinkage from 180 to 300°C.

Thermal expansion of the M90 and M110 concretes above 300°C therefore results from the combination of aggregate expansion and paste shrinkage, and also from cracking.

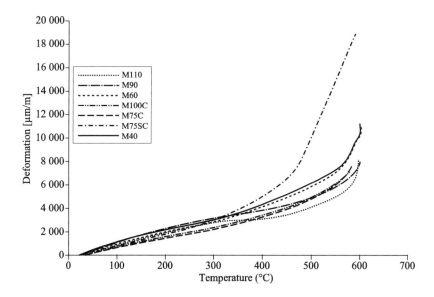

Figure 5.11. *Thermal expansion curves for SCCs [PIN 07] and HPCs [HAG 04]*

Figure 5.11 also shows the expansion curves obtained by Hager [HAG 04] with the same experimental apparatus on three HPCs (M75C: f_c=106.8 MPa, limestone aggregates, V_p=272 liters, W/C=0.38; M75SC: f_c=98.2 MPa, silico-calcium aggregates V_p=327 liters, W/C=0.33; M100C: f_c=112.8 MPa, limestone aggregates, V_p=276 liters, W/C=0.33).

Pineaud observed that the SCCs he tested deformed slightly more than the HPCs that Hager studied. On the other hand, at 400°C the expansion of M75SC is much higher than those of M75C and M100C. This is principally due to the presence of siliceous aggregates which cleave above this temperature.

5.3. Mechanical behavior of SCCs at high temperature

5.3.1. *Changes in compressive strength*

Table 5.5 summarizes the main studies on the mechanical behavior of self-compacting concretes at high temperature.

It is very difficult to determine mechanical behavior of concretes in fires or when subjected to high temperatures. For this reasons, most tests are conducted at controlled high temperatures or after recooling (residual tests).

Generally speaking, the change in compressive strength with temperature varies greatly from one study to another. It depends on several factors linked to the concrete composition (type and proportions of materials used), heating conditions (rate of temperature increase), test type (at high heat or after recooling), etc. However, several trends common to most of the studies can be discerned (Figure 5.12).

Source	Experimental conditions	High temperature behavior
[PER 04]	Heated at 4°C/min Temperature levels: 20, 200, 400, 600, 800°C High temperature and residual tests Initial strength: 40 to 88 MPa Cylindrical samples ϕ10x20 cm cubic samples 20x20x200 cm Study of 12 SCCs and 4 CCs containing limestone or glass fillers, or PPF	Continuous reduction in strength (to 800°C, strength is only 15% of its value at 20°C) Similar behaviors observed for SCC and CC except at 800°C, Zero residual strength at 800°C for SCCs with limestone fillers Significant spalling (for all SCCs without PPF)
[STE 04]	Fire ISO 4102 T.2 after keeping for 180 days 300 mm cube Initial strength: 25 to 75 MPa 8 SCCs studied	Residual strength at 700°C between 47 and 67% Correlation between increasing W/C and decreasing strength Significant spalling
[BOS 06]	ISO 834 fire test 12 SCCs studied, of which 7 contain PPF cubic samples 60x50x20 cm Strength from 35 to 54 MPa	Spalling observed in all SCCs without PPF
[NOU 06]	Temperature rise from 0 at a rate of 0.5°C/min up to 400°C Tests on SCCs with and without polypropylene fibers Strength of SCC without PPF is 81 MPa, 75 MPa for SCC with PPF	Residual strength of SCC with PPF: 41 MPa at 400°C (i.e. 55%) Explosive spalling for all SCCs without PPF

[ANN 07]	Heating at 3.5°C/min Temperature levels: 105, 250, 350 and 550°C Studies of an SCC and an CC, cubic sample 150 mm with initial strength 66 MPa for SCC and 57 MPa for CC	Reduction in relative residual strength reached 50% at 550°C for SCC and CC tested No instability observed Cracks observed in transition zones
[PIN 07]	High temperature test with temperature rise of 1°C/min Temperature levels: 120, 250, 400 and 600°C 9 SCCs studied Cylindrical samples ϕ10.4 x 30 cm Initial strength: 40 to 110 MPa	High temperature strength at 600°C between 35 and 60% Strength at 120°C lower than CC strength but strength increase at 250°C Increase in paste volume reduces the increase in compressive strength between 120 and 250°C W/B has a slight influence on changes in compressive strength at high temperature No instability observed
[SID 07]	Heating rate 5°C/min Temperature levels: 100, 300, 500 and 700°C 4 SCCs and 4 CCs studied Tests on cubic samples 150 mm with strengths 34, 43, 54 and 73 MPa kept in water	At 700°C, the relative residual resistance is 50 to 75% for SCCs Significant reduction in residual strength for CCs tested Explosive spalling observed for all SCCs and CCs with initial strength between 54 and 73 at temperatures between 380 and 458°C

[TAO 07]	Heating rate 5°C/min with pre-loading of 20% of the final load Temperature levels: 20, 200, 400, 600 and 800°C 2 SCCs with and without PPF studied Tests on cylindrical samples φ15x30 cm with initial strength: 70 MPa	At 800°C, 48% relative residual strength for SCCs with PPF Explosive spalling between 300 and 450°C for all SCCs without PPF
[YE 07]	Heating at 10°C/min to 950°C (keeping at ambient temperature, RH 60%) cubic samples 30x30x15 cm 2 SCC pastes and 1 HPC paste studied (PPF proportion: 0.5 and 1 kg/m^{-3})	Explosive spalling observed in SCCs without PPF
[LIU 08]	Heating at 10°C/min (keeping at ambient temperature, RH 60%) Temperature levels: 130, 200, 300, 400 and 500°C One SCC paste and one HPC paste studied (PPF proportion 0.5 and 1 kg/m^{-3})	Explosive spalling observed in SCCs without PPF when heated to 500°C No explosive spalling in HPCs with or without PPF
[FAR 09]	Heating at 1°C/min Temperature levels: 150, 300, 450 and 600°C Cylindrical samples φ16 x 32 cm with initial strength 37 to 54 MPa 2 SCCs and 1 CC studied	Reduction in strength between 20 and 600°C for the SCCs and CC tested Gain in strength observed at 300°C for SCCs Explosive spalling observed at 315°C for the 2 SCCs tested

Table 5.5. *Summary of research carried out on SCCs at high temperatures [STE 04, PER 04, FAR 09, ANN 07, TAO 07, SID 07, PIN 07, NOU 06, LIU 08, BOS 06].*

Between 100°C and 200°C, a reduction in compressive strength is observed in concretes which can reach 30% of the initial strength. Nevertheless, Persson [PER 04] observed better strength in SCCs between 100°C and 200°C relative to the ordinary HPCs tested. Between 250°C and 350°C, changes in the compressive strength are difficult to analyze: a reduction or an increase in strength may be observed. From Pineaud [PIN 07], the paste volume seems to play an important role in how the strength varies in this temperature range. The results from this author show, in effect, that an increase in paste volume leads to a reduction in the changes in strength.

The water/binder ratio, on the other hand, seems to play a more minor role, regardless of temperature. The strength increase in this temperature range has already been observed in conventional concretes [KAN 07, PHA 01, SAA 96, ZOL 01, MAL 89] and is therefore not specific to SCCs. Above 400°C, high temperature behavior of all SCCs, when the studies are brought together, is similar to that of conventional concretes: a decrease in compressive strength is observed and the gradient of this decrease is comparable for all concretes. No other difference has been observed between SCCs and conventional concretes. Khoury [KHO 92] attributes the reduction in the observed strength between 100°C and 200°C to the reduction in Van der Waals type bonding forces between flakes of C-S-H. Heating to such temperatures reduces the gel surface energy which develops in silanol groups (Si-OH·HO-Si) which give rise to weak bonding forces.

As seen in previous sections, exposure to high temperature also causes significant physico-chemical changes (some hydrates decompose, bonded water is lost) which modify the cement paste microstructure. From the work of Dias et al. [DIA 90], the loss of water, between 200°C and 300°C, leads to rehydration of the paste due to migration of water in the pores.

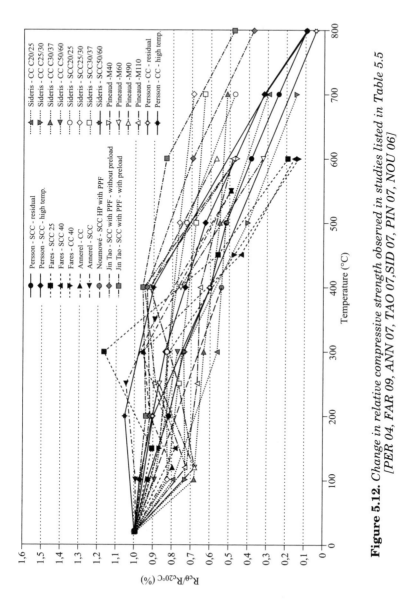

Figure 5.12. *Change in relative compressive strength observed in studies listed in Table 5.5 [PER 04, FAR 09, ANN 07, TAO 07, SID 07, PIN 07, NOU 06]*

For Khoury [KHO 92], the loss of a fraction of bonded water causes the formation of shorter and stronger bonds of siloxane compounds, with higher surface energies which contribute to shrinkage and increases in strength. Fares *et al.* [FAR 10] showed, with the aid of SEM image analysis, a complementary hydration of cement grains, which resulted from significant movements of water which occur during heating between 100 and 300°C. Dehydration of C-S-H causes water to be produced inside the cement paste as well as an increase in the porosity of these hydrates which allows the water easier access to the anhydrous cement grains.

In addition, since porosity increases for all concretes between 100 and 300°C by around 2 to 3%, the increase in concrete strength cannot be linked to the porosity filling up with new hydration products. The increase in strength observed can therefore be linked in part to the formation of hydration products which have better binding properties. This hypothesis agrees with the study carried out by Robert *et al.* [ROB 09]. These authors observed the formation of Katoite or Jaffeite between 150°C and 350°C which explains the compressive strength increase in concretes. This formation has better binding properties than C-S-H.

Above 350°C, compressive strength falls very quickly. Hydrate decomposition, particularly of portlandite and C-S-H, the appearance of cracking [NOU 95] and finally the allotropic transformation of quartz which weakens the aggregates, [FEL 00] explain this fall. At 800°C, the compressive strength of SCCs containing limestone fillers is effectively zero [PER 04], because of limestone decarbonation which takes place around 750°C.

5.3.2. *Elastic modulus*

Figure 5.13 shows the change in elastic modulus for different concretes (conventional concrete, HPC and SSC) as a function of temperature [BAM 07]. The changes are comparable for all concretes studied. At 300°C, a fall of around 40% in the elastic modulus is usually observed. Above 450°C, concrete rigidity is very low: a loss in the elastic modulus in the order of 70% is observed.

Fares *et al.* [FAR 09] and Pineaud [PIN 07] observed that the decrease in the elastic modulus for SCCs was slightly less than for conventional concretes. But Pineaud observed elsewhere that the increase in paste volume tends to speed up the decrease in the elastic modulus.

Modifications in concrete microstructure following heating are responsible for the fall in elastic modulus [GRO 73, LAB 74]. Damage to the matrix due to dehydration and cracking [HAG 04], cracking at the paste-aggregate interface [TOL 02] and degradation of the aggregates themselves above 300°C [BAM 07] are the principal factors which explain reductions in concrete rigidity.

5.4. Thermal stability

There is no standard test for studying high temperature stability of concretes. Several factors affect the results, such as the sample size, speed of temperature increase, heating mode, curing mode, saturation extent, and mechanical loading. The best of these options is to conduct tests on size 1 test pieces, but these tests are very costly.

To address this inconvenience, several authors used reduced-size samples. The increases in temperature vary. For SCCs, the majority of tests have been carried out using so-called "slow" heating with heating rates between 0.5 and 5°C/min.

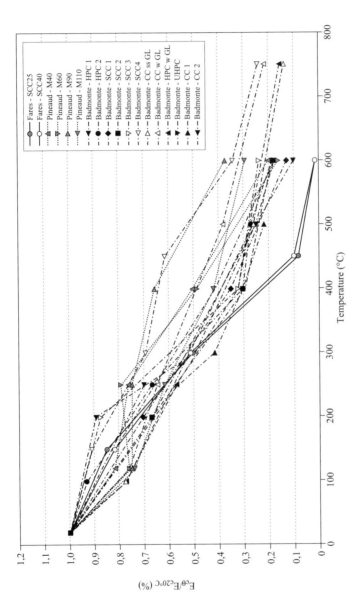

Figure 5.13. *Change in the elastic modulus with temperature [BAM 07, FAR 09, PIN 07]*

In the study by Fares *et al.* [FAR 09] (cylindrical samples 16x32 cm for two SCCs and one conventional concrete subjected to heating at 1°C/min to 150, 300, 450 and 600°C), two of the formulations showed thermal instability. When heated to temperatures higher than 300°C, the 2 SCC samples with 90 days compressive strengths of 37 and 54 MPa exploded violently. However, the explosive spalling only concerned cylindrical samples with dimensions φ16x32 cm and occurred around 315°C. As seen in Figure 5.14, pieces of the samples were scattered around the oven. Explosive spalling is random in character. In effect, all samples with the same mix, tested in exactly the same conditions, did not have the same behavior with regard to explosive spalling. SCCs seem more likely to explode than conventional concretes.

Figure 5.14. *Samples before and after heating [FAR 09]*

Persson [PER 04] studied the effect of the water/binder (W/B) ratio (from 0.40 to 0.70), mineral admixtures (limestone filler and glass filler) and polypropylene fiber content on SCC and conventional concrete thermal stability. Persson observed significant spalling as a function of the W/B ratio. When kept in water, SCC explosive spalling appeared when the W/B ratio was less than 0.40 and when kept in air, for a W/B ratio of less than 0.35. In creating conventional concretes in the same conservation conditions, with the same relative humidity, the conventional concretes had better explosive spalling resistance than the SCCs.

Sideris [SID 07] observed explosive spalling at W/B ratios of 0.45 and 0.46 for HP SCC (with strengths of 73 and 54 MPa respectively) and HPC (with strengths of 67 and 45 MPa and W/B of 0.43 and 0.46 respectively). In this study, SCCs had a higher residual strength than the conventional concretes. Explosive spalling was observed for all concretes tested.

Boström *et al.* [BOS 06] showed that limestone fillers affect SCC stability: increasing the filler proportion increases spalling in SCCs. Furthermore, the authors observed that a concrete without limestone filler which has an increased cement content seems to become more unstable than a concrete with limestone fillers and the same water/binder ratio.

Noumowé *et al.* [NOU 06] studied the thermal stability of HP SCCs (high performance SCCs) with and without polypropylene fibers with slow heating (0.5°C/min up to 400°C) and with quick heating (fire ISO 834 up to 600°C). Their observations are shown in Table 5.6.

Cylindrical samples (φ16 x 32cm)	Slow heating (0.5°C/min)	Fire ISO 834
HP SCC without PPF	Explosive spalling	Explosive spalling
HP SCC with PPF	No disturbance	No disturbance

Table 5.6. *Thermal stability of SCCs studied [NOU 06]*

The authors in [NOU 06] noticed that during slow heating, high performance SCCs showed instability at a temperature between 180°C and 250°C (Figure 5.15). By adding polypropylene fibers when the concretes are being mixed, the risk of explosive spalling is avoided. These tests confirm that polypropylene fibers improve thermal stability of SCCs and high performance SCCs.

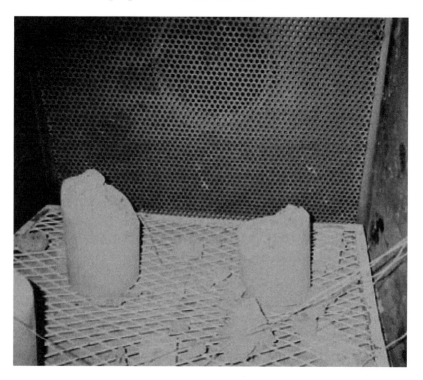

Figure 5.15. *Explosive breaking of SCC samples during an ISO 834 heating test [NOU 06]*

Ye *et al.* [YE 07] and Liu *et al.* [LIU 08] also studied the effect of polypropylene fibers on high temperature behavior of SCCs. The samples studied were kept for 28 days at ambient temperature and 60% relative humidity. Explosive spalling was observed in SCCs without PPF on all heated surfaces and for all samples of the same formulation and with the same type of heating. The addition of polypropylene fibers also reduced the risk of explosive spalling in SCCs to that of conventional concretes with a similar W/C ratio [YE 07].

Finally, by taking into account thermal instability risks, RILEM [RIL 07] recommends making the necessary arrangements when using SCCs (the use of thermal barriers or adding polypropylene fibers into the concrete).

5.5. Conclusion

All of the studies carried out internationally on the behavior of SCCs exposed to high temperatures lead to the following main conclusions:

– Transformations which occur inside SCCs and conventional concretes during heating are similar. The small difference observed result essentially from the presence of mineral admixtures (in SCC) and from different proportions of components in the mixture.

– As the temperature increases, porosity increases because of water loss and microcracking caused by differences in the expansion of the paste and aggregates. At temperatures less than 300°C, the change in microstructure (porosity and permeability) is linked to widening capillary pores due to the loss of water bonded to the pore network. At temperatures higher than 300°C, changes in transfer properties are linked to deterioration in the cement matrix which leads to modifications of the fine porosity in the concrete.

– Thermal properties in SCCs seem to be very close to those of HPC (conventional concrete). Thermal deformations are governed by aggregate expansion and paste shrinkage.

– As the temperature rises, the mechanical performance of SCCs is degraded. This reduction is linked to changes in the microstructure which is itself linked to dehydration and hydrate decomposition, as well as cracking.

– As the temperature rises, SCCs may show thermal instability. This instability may be characterized by spalling phenomena or violent explosive spalling of heated parts. To prevent this from happening, polypropylene fibers can be added to SCCs. At high temperatures, the fusing of these fibers affects the concrete's permeability, hence reducing the vapor pressure in pores and therefore the risk of thermal instability. However, at present, the only thermal stability tests that have been carried out have used only small or medium sized samples. Without a doubt, tests on a larger scale are needed in order to be able to draw any definite conclusions.

5.6. Bibliography

[ANN 07] ANNEREL E., TAERWE P., VANDEVELDE P., "Assessment of temperature increase and residual strength of SCC after fire exposure", *Proceedings of 5th International RILEM Symposium on Self-Compacting Concrete*, p. 715-720, Ghent, Belgium, 2007.

[BAM 07] BAMONTE P., FELICETTI R., "On the tensile behavior of thermally-damaged concrete", *Proceedings of 6th International Conference on Fracture Mechanics of Concrete and Concrete Structure*, FraMCoS6, vol. 3 Catania, Italy, 2007.

[BOS 06] BOSTRÖM L., JANSSON R., "Spalling of self-compacting concrete", *Proceedings of 4th International Workshop Structures in Fire*, vol. II, Aveiro, Portugal, 2006.

[DEJ 07] DEJONG M.J., ULM F.J., "The nanogranular behavior of C-S-H at elevated temperature (up to 700°C)", *Cement and Concrete Research*, vol. 37, no. 1, p. 1-12, 2007.

[DIA 90] DIAS W.P.S., KHOURY G.A., SULLIVAN P.J.E., "Mechanical properties of hardened cement paste exposed to temperatures up to 700°C", *ACI Material Journal*, no. 87, p. 160-166, 1990.

[DIV 05] DIVET L., *Présentation des techniques de diagnostic de l'état d'un béton soumis à un incendie*, Techniques et méthodes des Laboratoires des Ponts et Chaussées (col.), Presses des Laboratoires des Ponts et Chaussées, Paris, 2005.

[FAR 09] FARES H., NOUMOWÉ A., RÉMOND S., "Self-compacting concrete subjected to high temperature: mechanical and physico-chemical properties", *Cement and Concrete Research*, vol. 39, no. 12, p. 1230-1238, 2009.

[FAR 10] FARES H., RÉMOND S., NOUMOWÉ A., COUSTURE A., "High temperature behaviour of self-consolidating concrete: microstructure and physico-chemical properties", *Cement and Concrete Research*, vol. 40, no. 3, p. 488-496, 2010.

[FEL 00] FELICETTI R., GAMBAROVA P.G., SORA M.N., KHOURY G.A., "Mechanical behaviour of HPC and UHPC in direct tension at high temperature and after cooling", *Fifth RILEM Symposium on Fibre-reinforced Concretes*, p. 749-758, Lyon, 2000.

[GAL 01] GALLÉ C., SERCOMBE J., "Permeability and pore structure evolution of silico-calcareous and hematite high-strength concretes submitted to high temperatures", *Material and Structure*, vol. 34, p. 619-628, 2001.

[GRO 73] GROSS H., "On high temperature creep of concrete", *International Conference on Structural Mechanics in reactor Technology – 2nd SMIRT*, West Berlin, vol. 3, 1973.

[HAG 04] GAWESKA HAGER I., Comportement à haute
température des bétons à haute performance, PhD Thesis,
Ecole Nationale des Ponts et Chaussées – Ecole
Polytechnique de Cracovie, 2004.

[KAN 07] KANEMA M., Influence des paramètres de
formulation et microstructuraux sur le comportement à
haute température des bétons, PhD Thesis, University of
Cergy-Pontoise, 2007.

[KHO 92] KHOURY G.A., "Compressive strength of concrete
at high temperature: a reassessment", *Magazine of
Concrete Research*, no. 161, p. 291-309, 1992.

[KHO 95] KHOURY G.A., "Strain components of nuclear-
reactor-type concretes during first heat cycle", *Nuclear
Engineering and Design*, no. 156, p. 313-321, 1995.

[JAN 04] JANSSON R., Measurement of thermal properties at
elevated temperatures – Brandforsk Project 328-031, SP
Report, no. 46, 2004.

[JUM 89] JUMPPANEN U.M., "Effect of strength on fire
behavior of concrete", *Nordic Concrete Research*, no. 8,
p. 116-127, 1989.

[LAB 74] LABANI J.M., SULLIVAN P.J.M., The performance of
lightweight aggregate concrete at elevated temperature,
Imperial College: concrete structure and technology,
Reports CSTR, no. 73/2, 1974.

[LIU 06] LIU X., Microstructural investigation of self-
compacting concrete and high performance concrete
during hydration and after exposure to high temperature,
PhD Thesis, University of Ghent, Belgium, 2006.

[LIU 08] LIU X., YE G., DE SCHUTTER G., YUAN Y., TAERWE
L., "On the mechanism of polypropylene fibres in
preventing spalling in self-compacting and high-
performance cement paste", *Cement and Concrete
Research*, vol. 38, p. 487-499, 2008.

[MAL 89] MALHOTRA V.M., WILSON H.S., PAINTER K.E., "Performance of gravel stone concrete incorporating silica fume at elevate temperatures", *Proceedings of Trondheim Conference*, p. 1051-1076, Trondheim, Norway, 1989.

[NON 99] NONNET E., LEQUEUX N., BOCH P., "Elastic properties of high alumina cement castables from room temperature to 1600°C", *Journal of the European Ceramic Society*, vol. 19, p. 1575-1583, 1999

[NOU 95] NOUMOWE A., Effet des hautes températures sur le béton: cas particulier des BHP, doctoral thesis, INSA de Lyon, 1995.

[NOU 06] NOUMOWÉ A., CARRÉ H., DAOUD A., TOUTANJI H., "High strength self-compacting concrete exposed to fire test", *Journal of Materials in Civil Engineering*, ASCE, vol. 18, no. 6, p. 754-758, 2006.

[PER 04] PERSSON B., "Fire resistance of self-compacting concrete", *Materials and Structures*, vol. 37, p. 575-584, 2004.

[PER 86] PERSY J.P., DELOYE F.X., "Investigations sur un ouvrage en béton incendié", *Bulletin des laboratoires des Ponts et Chaussées*, vol. 145, p. 108-114, 1986.

[PHA 01] PHAN L.T., LAWSON J.R., DAVIS F.L., "Effects of elevated temperature exposure on heating characteristics, spalling and residual properties of high performance concrete", *Materials and Structures*, vol. 34, p. 83-91, 2001.

[PIM 01] PIMIENTA P., "Propriétés des BHP à hautes températures – Etude bibliographique", *Cahier du CSTB*, no. 3352, 2001.

[PIN 07] PINEAUD A., Contribution à l'étude des caractéristiques mécaniques des bétons auto-plaçants et application à l'industrie de la préfabrication, PhD Thesis, University of Cergy-Pontoise, 2007.

[PLA 02] PLATRET G., "Suivi de l'hydratation du ciment et de l'évolution des phases solides dans les bétons par analyse thermique, caractéristiques microstructurales et propriétés relatives à la durabilité des bétons", *Méthodes de mesure et d'essai de laboratoire,* Méthode d'essai n°58, Laboratoire Central des Ponts et Chaussées, Paris, 2002.

[RIC 99] RICHARD N., Structure et propriétés élastiques des phases cimentières à base de mono-aluminate de calcium, PhD Thesis, University of Paris VI, 1999.

[RIL 07] RILEM, Durability of self-compacting concrete, RILEM, Report 38, 2007.

[ROB 09] ROBERT F., VÉRON E., MORANVILLE M., MATZEN G., "Link between cement paste chemical changes and mortar mechanical resistance at high temperature", *First International RILEM Workshop on Concrete Spalling Due to Fire Exposure, Proceedings*, Leipzig, Germany, 2009.

[SAA 96] SAAD M., ABO-EL-ENEIN S.A., HANNA G.B., KOTKATA M.F., "Effects of temperatures on physical and mechanical properties of concrete containing silica fume", *Cement and Concrete Research*, vol. 26, p. 669-675, 1996.

[SHA 99] SHA W., O'NEILL E.A., GUO Z., "Differential scanning study of ordinary Portland cement", *Cement and Concrete Research*, vol. 29, p. 1487-1489, 1999.

[SID 07] SIDERIS K., "Mechanical characteristics of self-consolidating concretes exposed to elevated temperatures", *Journal of Materials in Civil Engineering*, ASCE, vol. 19, no. 8, p. 648-654, 2007.

[STE 04] STEGMAIER M., REINHARDT H.W., "Fire behaviour of plain self-compacting concrete", *Otto-Graf-Journal*, vol. 15, p. 33-42, 2004.

[TAO 07] TAO J., LIU X., YUAN Y., "High strength self-compacting concrete at elevated temperature", *Proceedings of the 5th International RILEM Symposium on Self-Compacting Concrete SCC 2007*, no. 54, p. 1135-1144, Ghent, Belgium, 2007.

[TOL 02] TOLENTINO E., LAMEIRAS A.M., GOMES F.S., RIGO DA SILVA C.A., VASCONCELOS W.L., "Effects of high temperature on residual performance of Portland cement concrete", *Materials Research*, vol. 5, no. 5, p. 301-307, 2002.

[TSI 97] TSIMBROVSKA M., KALIFA P., QUENARD D., "High performance concrete at elevated temperature: permeability and microstructure", *SMIRT 14*, Lyon, 1997.

[YE 07] YE G., DE SCHUTTER G., TAERWE L., "Spalling behaviour of small self-compacting concrete slabs under standard fire conditions", *Proceedings of the 5th International RILEM Symposium on Self-Compacting Concrete SCC 2007*, no. 54, p. 799-804, Ghent, Belgium, 2007.

[ZOL 01] ZOLDNER N.G., "Effect of high temperatures on concretes incorporating different aggregates", *Proceeding ASTM*, vol. 30, p. 1087-1108, Philadelphia, United States, 2001.

Glossary

CC: Conventional concrete

HPC: High performance concrete

HP SCC: High performance self-compacting concrete

PPF: Polypropylene fibers

SCC: Self-compacting concrete

List of Authors

Sofiane AMZIANE
Blaise Pascal University
Clermont-Ferrand
France

Geert DE SCHUTTER
University of Ghent
Belgium

Hana FARES
IUT Nancy-Brabois
University of Nancy
France

Abdelhafid KHELIDJ
University of Nantes
France

Christophe LANOS
University of Rennes 1
France

Ahmed LOUKILI
Ecole Centrale Nantes
France

Michel MOURET
Paul Sabatier University
Toulouse
France

Albert NOUMOWÉ
University of Cergy-Pontoise
France

Sébastien RÉMOND
Ecole des Mines de Douai
France

Emmanuel ROZIÈRE
Ecole Centrale Nantes
France

Stéphanie STAQUET
Free University of Brussels
Belgium

Philippe TURCRY
University of Rochelle
France

Thierry VIDAL
Institut National des Sciences Appliquées
Toulouse
France

Index